# Jesus:

# Lord
## of
# Time and Space

# Jesus:

# Lord
# of
# Time and Space

## Lambert T. Dolphin

**Front Cover:**

The Dumbell Nebula in Vulpecula
(Messier 27, NGC 6853)
This planetary nebula is about 3 light-years,
 or $2 \times 10^{13}$ miles in diameter.

New Leaf Press

**First Edition
1988**

Cover:
Lick Observatory photograph
X-16: Messier 27, NGC 6853
Used with permission

Library of Congress Catalog Number: 88-60411
ISBN: 0-89221-151-2

# Table of Contents

Acknowledgments..................................................6

Introduction..................................................7

One.  The Heavens and the Earth........................13

Two.  Science and Revelation.............................21

Three.  Is the Universe Running Down?...............39

Four.  The Steps of Creation............................61

Five.  The Complexities of Time.........................85

Six.  Is Light Slowing Down?............................105

Seven.  Changing "Constants" in Physics?............131

Eight.  Flawed Men (and Women)........................139

Nine.  Jesus, Lord of Angels...........................171

Ten.  Physics in a Fallen Universe...................199

Eleven.  The Undoing of Cosmic Evil..................227

Twelve.  Jerusalem Above, The Mother of us All.........251

Appendix.  What the Bible Says About Itself...........261

Footnotes are Included at the End of Each Chapter

# Acknowledgments

I am especially glad that several loyal friends took the time to read and comment most helpfully on the manuscript. My thanks to Trevor Norman, Barry Setterfield, Elana Lynse, Ron Cooper, Nancy Del Grande, David Lauben, and Bryce Self.

Since I would never written a manscript in the first place without much encouragement from my friend of nearly twenty years, Tim Spacek, he deserves special thanks and recognition.

Mike McGinnis, a Graphics Design student at San Jose State University undertook to turn my MacPaint diagrams into more artistic final drawings. Carl Gallivan and Dan Ledwick offered me helpful suggestions with fonts and formatting when I elected to do my own "typesetting" on a MacIntosh. Any formatting errors are therefore mine not my faithful publishers!

Clifford Dudley and his staff at New Leaf Press have encouraged and helped me greatly in the editing and publishing of this book. I appreciate the introduction my friend Dr. David A. Lewis gave me to Mr. Dudley.

This book is dedicated to my many friends in the younger generation who are seeking for and following Jesus Christ, **"At one time we too were foolish, disobedient, deceived and enslaved by all kinds of passions and pleasures. We lived in malice and envy, being hated and hating one another. But when the kindness and love of God our Savior appeared, he saved us, not because of righteous things we had done, but because of his mercy. He saved us through the washing of rebirth and renewal by the Holy Spirit, whom he poured out on us generously through Jesus Christ our Savior, so that, having been justified by his grace, we might become heirs having the hope of eternal life,"** (Titus 3:3-7) ( NIV).

# Introduction

Without apology, this book is an open statement of my personal faith and commitment to Jesus Christ. I have been a Christian only about half my life, since the age of 30. I rebelled against the nominal Christianity I was exposed to as a boy, became enamored with science, and soon took the position of an atheist, antagonistic to such "old fashioned" and out of date ideas as those held by my godly grandmother Alma L. Dolphin of Emmett, Idaho. She, however, continued to pray faithfully for me for thirty years and by her life demonstrated more selfless concern for others than I have encountered since. My anti-Christian stance was based of course on my ignorance of the actual content of Christianity and the arrogance of my intellectual pride.

My entire pre-Christian life was characterized by an intense and insatiable quest for "truth", and I was quickly dissatisfied with one philosophy after another as I came to see that none offered any inner knowing that transformed human experience. I wanted something to fill what seemed to be a dark, hollow void in my innermost self. My mother's death when I was 14 haunted me with the thoughts of my own mortality and told me of the all-too-real disasters in life for which no one I could knew had any satisfying answers.

I am very grateful that my father sacrificially paid my way through college at San Diego State University and two years of graduate work in physics at Stanford University. I shall never forget the great men of physics, including several marvelous Nobel Prize winners, I sat under at Stanford. To this day I highly admire them, their knowledge and accomplishments, and I am most thankful to God for allowing me such a privileged education in the highest courts of secular science. Yet my graduate school experience left me with the sense that I lived in a meaningless, cold, impersonal universe. Everything had happened by accident apparently, and no one spoke of man's place in the universe or of the possibility of a Creator. I came to believe that physics would not satisfy my soul. I was at a low point in my search when a "summer" job at Stanford Research Institute (now SRI

International) in 1956 provided a permanent diversion from the completion of my PhD.

My subsequent career - at first, largely devoted to the physics of the upper atmosphere and later, to remote sensing in geophysics of the earth itself - has been most rewarding and challenging indeed. Presently, I am a private consultant in physics, which gives me more time to devote to Christian ministry and to long-delayed writing projects such as this book.

At the suggestion of a brilliant friend in Medical School at Stanford who was pursuing psychiatry, I decided - about 1957 I suppose, that a controlled search through the inner space of my own unconscious might be the best way to resume my quest for the meaning of life. I sought out the best classical Freudian psychiatrist to work with and devoted three days a week to this new "educational study" for nearly three years. It was not because I felt myself to be neurotic or non-functioning in the ordinary sense. Rather, a restless hunger for truth compelled me towards this new avenue of personal "research" into life. Freud's outspoken atheism led me to Carl Jung's positive assessment of religious experience. My very rational, all-too-logical mind was challenged as I saw the value of emotions and the reality that dreams could convey about the state of one's own inner world of the unconscious. Many of my adult life problems could be traced back to childhood experiences, as Freud had suggested, and yet I wondered when I would change and begin to be at peace with myself, finding the all-elusive happiness I assumed almost everyone else had attained. I also wondered what my psychiatrist talked to his psychiatrist about! Had he found "the answer"? I experimented with LSD and mescaline under controlled, legal, medical conditions and became convinced of the existence of spiritual realities. (I paid for that foolish experience with a psychosis that took the better part of a year to recover from!)

With new enthusiasm I then sought the best astrologer I could find, to be certain I had not missed any esoteric insights about myself, and I eagerly followed the eloquent teaching of Bay Area resident Alan Watts. Watts, a former Episcopal priest, left the Christian faith to become a spokesman for eastern philosophy, eastern religion, and back-to-nature lifestyle. His books and teaching were lucid, attractive and exciting. But one day in attending one of his Sausalito houseboat lectures, I found he could not answer my question about why the Bible apparently spoke of two resurrections from the dead rather than simple reincarnation

8

as a description of the events that follow physical death. Watts seemed unable to communicate his early Christian views. I had read one of his early books on Christianity and was quite drawn to what it had to say, so I wondered why he could not, or would not, speak of his former western "faith."

At the age of thirty I thought I had exhausted all the possibilities. I was drinking too much. I felt that no permanent happiness or love existed in the universe. I suspected there were probably no moral absolutes at all. Suicide seemed a good way out - except for the remote possibility that there might be a hell. Suicide, I knew, was irreversible, and I was after all fairly cowardly about risks of that order. Why not wait a bit longer?

I am certain it was the answer to the prayers of my faithful grandmother that led me to accept the invitation of my current roommate's parents, Verle and Marjorie Thomson, to attend church in 1962. There I realized that though I had read widely about the religions of the world, I never had opened a Bible! It seemed to me utterly amazing that I could have passed through public schools, college, and much professional life without ever learning anything about the "best known" and "most popular" book known to man! I then turned all my questing energies towards the Bible, looking for flaws, inconsistencies and errors. My curiosity was aroused, and I was suspicious that there might be some sort of invisible cosmic conspiracy operating in society that actually operates to keep people from reading the Bible! I was upset and disturbed by notions of my manifold "sins," but Jesus held an immense attraction for me by what He said and by the way He lived. Within a few months I discovered, in Pastor Ralph Kraft's office, that a personal commitment to Jesus as Lord brings an absolutely clear (if not immediate) response from God. In my case, my decision for Christ brought about such a sudden and total change in my consciousness that I knew I had been reborn and remade all in one moment.

My Christian faith expanded my interest in science rather than "narrowing my mind." At first I felt I should head off for the seminary. Fortunately for me, both the seminary officials and my pastor knew what was better for me, so I began learning the Bible on my own and never regretted being Spirit-taught and built up in two vital Silicon Valley churches where plenty of scholarship and sound theology abounds on all sides. Soon I was teaching the Bible and speaking to large audiences about my conversion. I received great encouragement and support from

9

Mary Clark and Helen Baugh of Stonecroft Ministries who first published the account of my conversion in a small booklet "My Search."

I have made a few feeble attempts at writing short books and articles in the past but, until late in 1987, lacked time to devote to a book such as this. The informed Christian reader will find that most of my biblical ideas are not my own but have come from my teacher of many years, Bible Scholar and Expositor, Ray C. Stedman of Peninsula Bible Church, Palo Alto, where I am now a member. I have been greatly influenced by C. S. Lewis, and blessed by the late Canadian scholar Arthur C. Custance. I have added dozens of books about science and on the Bible to my library. I have felt, however, that more needed to be said about the relationship between science and the Bible.

Recent exciting work by Australian scientists Trevor Norman and Barry Setterfield, suggesting that some of the "constants" of physics have changed with time, provided me with the impetus to actually sit down to write this book.

My first concern as a scientist writing on this subject, **"Jesus, Lord of Time and Space"** is that I believe science has come to deal too exclusively and authoritatively with the physical world, forgetting both the inherent limits of science and the Biblical revelation that the invisible world of the spiritual is the more important half of creation.

I have tried to help the layman understand, in Chapter Two, that science does indeed work within rather narrow limitations and that most of the really important things in life come to our senses by revelation from God. This revelation is given to us in nature, within our inner selves, in the Bible, and supremely in God's Son, our Lord Jesus Christ. We have received more revelation than most of us care to admit, and for all of it, God holds us accountable!

To introduce the non-scientist to the world of science that I am very familiar with, I felt an easy transition might be to address the notion (Chapter Three) that the universe is running down. Mathematically and philosophically, the concept of entropy (tendency of the universe towards inertness) is a profound subject. My purpose is to show very simply that the universe we live in is, in fact, becoming more disorderly and chaotic, and that energy is becoming less and less available. In the physical world things are not improving, but decaying away, exactly as scripture

10

says. The universe, as science knows it, is not self-renewing and not progressing to states of high order and evolutionary progress!

Because I value the book of Genesis in my life, having been shown, as a new Christian by men and women of experience, that the Bible is indeed accurate, authoritative and dependable, I have devoted Chapter Four to a brief discussion intended to show that God's activity in creation takes place largely from outside of time, and that even a casual reading of Genesis will show that such notions as the "Big Bang" theory for "creation" are naive, (to say the least).

Many of my creationist brothers-in-Christ do not seem, in my opinion, to understand the nature of time as revealed in scripture. So based on what I have learned from the scholarly Biblical studies of Arthur Custance, I have attempted in Chapter Five to show the reader that time and eternity are vast and complex subjects that can not be reduced to our ordinary ideas of clock time as measured in our daily lives today.

In Chapters Six and Seven, I introduce the reader to the seven-year research study of two Australian colleagues, physicist-astronomer Barry Setterfield and mathematician Trevor Norman. I consider their report on the possibility that the speed of light has not been constant to be a monumental and careful effort. They may well have unloosed a major revolution in physics. I suspect they have, but time will tell. If I read my Bible correctly, God will soon upset the status quo rather radically in any case!

Knowing both from contemporary science and from the Bible that man somehow has a central place in the universe, I have felt it appropriate to discuss, in Chapter Eight, the notion that our universe, as we know it today, is seriously flawed. Man is also flawed and has lost the power to perform his assigned task of stewardship over creation on God's behalf. Nature has become subjected to decay that was not present before man fell. Active, destructive evil exists both in the material and in the spiritual realms. The original creation has been seriously damaged by these powerful malevolent forces, and quite likely the laws of physics have been affected as well.

To draw a bit more attention to the vast realm of the spiritual world, I have commented briefly on the role Jesus has over the angels and the place the angels have in the government and regulation of the universe. This is the subject of Chapter Nine. In writing this book I must admit that I have raised questions in my own mind about the subject of miracles and the role of angels in creation that leave me more unsatisfied than ever about this complex subject! Chapter Ten presents my view of the geophysical history of creation in brief, taking into account the hypothesis that the velocity of light has decreased with time.

Finally, to avoid leaving the reader despondent and hopeless, I have attempted to show that the deadly evil infesting the entire old creation has, in fact, already been completely and totally resolved and undone by the death of Jesus Christ on the Cross. This is the subject of Chapter Eleven. What Christ did in six hours on earth is actually but a intrusion into our limited space-time frame of an eternal event, and an eternal transaction involving all the Persons of the Godhead. As a result of the death and resurrection of Jesus, God has become free to set about building an entirely new creation, a new race of men, and a new universe. The unveiling of the new will break forth in splendor and completeness at the very hour the old world order and the old creation suffers violent and total destruction, (Chapter Twelve). I do not hesitate to proclaim my firm belief that history is headed toward the consummation of all things in the personal, bodily return of Jesus to the Mount of Olives in Jerusalem, in literal fulfillment of what was declared long ago. Surely that time and that hour cannot be delayed much longer!

In this book I am not setting forth any new scientific dogmas I wish my readers to subscribe to and support. My intent has been to be provocative and to acquaint the non-scientist with the actual fact that within the world of science there are many things vague and nebulous that rest on uncertain ground and will surely change with time. I am more sure of my theology and hope that this book in some small way my excite the reader to take up the challenge Jesus left us when He said, "I have come that you might have life and have it more abundantly."

*Lambert Dolphin,*                                        *Cupertino, California*

12

**Chapter One**

## "...The Heavens and the Earth"

### Two Realms of Creation

I'd like to begin, not with a commentary on the Genesis account of creation, but by highlighting a point often missed by scientist and theologian alike when discussing the nature of the universe as we know it.

What I want to call attention to is that God created the "heavens *and* the earth." A cursory reading would suggest that this passage refers to the planet earth and the space beyond its atmosphere - encompassing moon, planets, sun, stars, and galaxies. The Bible could just as well have opened by stating that God created everything that exists and left it there. But there is a reason for distinguishing between the heavens and the earth. The former term evidently includes more than the clouds, the upper atmosphere, and space. It includes the spiritual dimension, whereas the latter includes the physical earth, but in more general terms - the material universe. Because we can neither see nor measure the spiritual world, we all must learn to realize that it is, nonetheless, real and greatly important.

The opening verse of Genesis tells us, of course, that the universe had a definite beginning, that it was brought into existence by God (*Elohim*) and was not manufactured out of pre-existing materials or born out of an act of copulation by the gods of mythology. Creation *Ex Nihilo* means creation out of nothing by a sovereign act of God. This is the Biblical view.

A notion commonly held by many is that the Hebrews of old believed in (at least) a "three-story" universe: the realm under the earth (the underworld); the earth itself housing mankind and society; and the heavens above including the lower atmosphere, space, and, last of all, at the topmost level, the throne of God. As far as I know, the point of view today among religious Jews is not especially three-tiered or transcendent, but more integrated - with matter and spirit treated as a whole creation. (Indeed, the Jews readily treat material objects and places as holy or unclean as easily as Christians regard only spiritual things as holy and capable of defilement.) The Biblical view of the universe is not one that is scaled in the vertical direction, with hell at the bottom, beneath the earth, and heaven at the top inaccessibly high above the reach of space ships and telescopes alike. To view heaven as "up" only means that heavenly things transcend the physical, and that the Creator is greater than the creation:

**"For my thoughts are not your thoughts, neither are your ways my ways, says the LORD. For as the heavens are higher than the earth, so are my ways higher than your ways and my thoughts higher than your thoughts."** (Isaiah 55:8-9 NAS)

Isaiah also says that God is **"the High and Lofty One who inhabits eternity..."** (Isaiah 57:15) Such statements emphasize that God stands apart from His creation and is awesome in His majesty and holiness. The transcendent greatness of our God surely needs rediscovering in our age of spiritual mediocrity and indifference to the Living Spirit who brought everything into existence from beyond time and space. However, many other statements in scripture call attention to God's intimate Presence everywhere. My main point is this: there are two parts to our universe, two aspects of creation. One is material and the other spiritual. One is discoverable by the five senses of man; the other is accessible to us by divine revelation and by experience - for man himself is both spirit and matter.

## Immersed in the Spiritual

The spiritual is not, however, far from the earth and outside of space and time beyond the stars. It surrounds us within and without. In fact we are *immersed* in spirit, and God Himself is a Spirit. When Paul the Apostle visited Athens for the first time and spoke to the citizens of that great city, he noted:

**"Men of Athens, I perceive that in every way you are very religious. For as I passed along, and observed the objects of your worship, I found an altar with this inscription, 'To an unknown God.' What you therefore worship as unknown, this I proclaim to you. The God who made the world and everything in it, being Lord of *heaven and earth*, does not live in shrines made by man, nor is he served by human hands, as though he needed anything,**

14

since he himself gives to all men life and breath and everything. And he made from one (man, Adam) every nation of men to live on all the face of the earth, having determined allotted periods and the boundaries of their habitation, that they should seek God, in the hopes that they might feel after him and find him. *Yet he is not far from each one of us, for 'In him we live and move and have our being*;' as even one of your poets have said, 'For we are indeed his offspring.' Being then God's offspring, we ought not to think that the Deity is like gold, or silver, or stone, a representation by the art and imagination of man. The times of ignorance God overlooked, but now he commands all men everywhere to repent, because he has fixed a day on which he will judge the world in righteousness by a man whom he has appointed (Jesus), and of this he has given assurance to all men by raising him from the dead." (Acts 17:22-31)

In the dimension where God himself dwells there are splendors beyond all human comprehension. Yet no man has seen God (the Father) at any time; all our glimpses are indirect. Paul wrote to young Timothy,

"I charge you to keep the commandment unstained and free from reproach until the appearing of our Lord Jesus Christ; and this will be made manifest at the proper time by the blessed and only Sovereign, the King of kings and Lord of lords, who alone has immortality and dwells in unapproachable light, whom no man has seen or ever can see. To him be honor and eternal dominion. Amen." (I Timothy 6:14-16)

The spiritual realm, which lies behind the smallest atomic particles, within the innermost part of man, and beyond the galaxies is commonly referred to as heaven, the heavenlies, or the heavenly places in the Bible. (The Bible does not use the term "supernatural.") Heaven is the realm where angels dwell. It is the world of the source of all things, the dimension of the permanent, the eternal, the enduring: **"Now faith is the assurance** (*hupostasis* = "to stand under", i.e., support, foundation) **of things not seen. For by faith the men of old gained divine approval. By faith we understand that the world** (*aionos* = ages, or world) **was created** (*katartizo* = to fit, or render complete) **by the word** (*rhemati* = the oracles, sayings, or spoken utterances) **of God, so that what is seen came into being out of that which is unseen."** (Hebrews 11:3  My paraphrase)

The physical world, the material realm, is perfectly real and solid (not *maya*, or illusion, as Hinduism supposes), but it is the

15

world of the fading, the transitory, the impermanent, and the perishable. This was not necessarily the way the universe was *created*, but it is the way things are now! Evil has disturbed our universe, interfered with both the realm of the spirit and the laws of physics. Evil (both angelic and human) has destroyed the original close and harmonious coupling between the spiritual and material dimensions of existence. What we now see and observe and experience is not the creation as it was finished at the end of the sixth day, but an aging "old" creation. If we choose to know God through faith in Jesus His Son, we perceive also that we are being made part of a new race, and prepared to live in a new creation which is now under construction:

**"So we do not lose heart. Though our outer nature is wasting away, our inner nature is being renewed every day. For this slight momentary affliction is preparing for us an eternal weight of glory beyond all comparison, because we look not to the things that are seen but to the things that are unseen; for the things that are seen are transient, but the things that are unseen are eternal."** (II Corinthians 5:6-8)

## The Spiritual World is the Source of All Things

As I hope to show, mankind's preoccupation through the centuries has been with the physical world, which in reality is the world of derived things and shadows. The spiritual world, on the other hand, is the solid, permanent and more splendid - not the other way around. But we find ourselves habitually looking at what has been derived, not at the Source.

A number of references in scripture tell us that things built by God in the spiritual world are more solid, permanent, and durable than their "shadowy" and temporary counterparts in the physical world. For example, while on Mt. Sinai, God told Moses to erect a Tabernacle and equip it with an elaborate set of furnishings: an altar, a laver, a great lampstand, a table of incense, a table for the shewbread, the Ark of the Covenant. The ark had to be built exactly as prescribed in every detail, **"...And see that you** (Moses) **make them** (all these things) **after the pattern for them which is being shown to you on the mountain."** (Exodus 25:40)

The writer of the letter to the Hebrews in the New Testament mentions the heavenly tabernacle when referring to Jesus as our Great High Priest:

16

"Now the point in what we are saying is this: we have such a high priest, one who is seated at the right hand of the throne of the Majesty in heaven, a minister in the sanctuary and the true tabernacle which is set up, not by man, but by the Lord." (Hebrews 8:1-2) The writer continues:

"But when Christ appeared as a high priest of the good things that have come, then through the greater and more perfect tent, (not made with hands, that, is not of this creation) he entered once for all into the Holy Place, taking not the blood of goats and calves but his own blood, thus securing an eternal redemption...under the law (of Moses) almost everything is purified with blood, and without the shedding of blood there is no forgiveness of sins. Thus it was necessary for the copies of the heavenly things to be purified by these rites, but the heavenly things with better sacrifices than these. For Christ has entered, not into a sanctuary made with hands, *a copy of the true one,* but into heaven itself, now to appear in the presence of God on our behalf." (Hebrews 9:11-12, 22-24). Yet another reference to this heavenly tabernacle or temple is given in Revelation 15:5-8.

One cannot hope to learn much about reality by looking at the shadows of things instead of their real form and substance! We cannot hope to understand ourselves or the universe (the heavens and the earth) if we ignore the information God has given us about the whole package.

## The Man Who Runs the Universe

We live in a *uni*verse, a moral, integrated creation. In spite of its present flaws, it is understandable, organized, and presently under the complete control of one man, Jesus of Nazareth, the Son of God. He is raised from the dead and has been placed, now, in the place of supreme authority over both the old creation and the new. Jesus is not now hanging on a cross outside of Jerusalem, nor is His body still in its borrowed grave. He is alive and has been elevated to the highest position of preeminence in the universe, with all authority and power having been given to Him by His Father. **Jesus, Lord of the Time and Space,** is in charge of the heavens and the earth, and it is He who manages every detail in both realms of creation:

"...I turned to see the voice that was speaking to me, and on turning I saw seven golden lampstands, and in the midst of the lampstands one like a son of man, clothed with a long robe and with a golden girdle round his breast; his head and his hair were white as white wool, white as snow; his eyes were like a flame of fire, his feet were like burnished bronze, refined as in a furnace, and his voice was like the sound of many waters; in his right hand he held seven stars, from his mouth issued a sharp two-edged sword, and his face was like the sun shining in full strength. When I saw him, I fell at his feet as though dead. But he laid his right hand upon me, saying, 'Fear not, I am the first and the last, and the living one; I died, and behold I am alive for evermore, and I have the keys of Death and Hades.'" (Revelation, 1:12-18).

## The Universe, A Home for Man

The Bible says we are not alone in the universe, that we need not be afraid of what we have not yet seen or experienced, either out among the stars, or in the heavenly places of the spiritual dimension. During the last week of his life, when so much happened in the life of Jesus that large portions of the gospels concentrate on those 7 or 8 days, Jesus gathered His disciples for the last time and spoke many comforting words to them in what is now called "The Upper Room Discourse." (John 14-16) One section of this discourse is especially relevant to our understanding the realms of creation, spiritual and material, and the nature of time:

Jesus said, "Let not your hearts be troubled; believe in God, believe also in me. In my Father's house are many rooms; if it were not so, would I have told you that I go to prepare a place for you? And when I go and prepare a place for you, I will come again and will take you to myself, that where I am you may be also." (John 14:1-3)

Here "the Father's house" is clearly the universe we live in, the whole universe, which is made up of both the visible and the invisible. The universe, Jesus implied, is to be compared to a house having many rooms, all made to live in. Like the homes[1] we live in, the various rooms serve various purposes. We have living rooms, dining rooms, bedrooms and perhaps a library-study.

The first helpful thing about this passage is that it teaches us that heaven is a better and more pleasant home than the best we know

18

here. This is not fiction or myth; Jesus was describing the way things really are. Further clues about the universe as a "house" designed to be lived in can be found elsewhere in the Bible. For example Jehovah, (Yahweh), says in Isaiah:

"For thus says the LORD, who created the heavens (he is God!), who formed the earth and made it (he established it); he did not create it a chaos, he formed it (to be inhabited!): I am the LORD, and there is no other." (Isaiah 45:18)

"Heaven is my throne and the earth is my footstool; what is the house which you would build for me, and what is the place of my rest? All these things my hand has made, and so all these things are mine, says the LORD. But this is the man to whom I will look, he that is humble and contrite in spirit and who trembles at my word." (Isaiah 66:1,2)

If we know now only that God's house contains a living room chair and its accompanying footstool, we nevertheless can infer the existence of kitchens, closets, banquet halls and libraries. For example, God's library must surely contain books in four-dimensional living color that contain in minute detail the history of the world as it really happened. Surely we shall find video tapes there containing the lives of all who have ever lived with thoughts, motives, and actual facts in open-books before us! After tiring of the library, we might like to move outdoors and investigate the gardens in heaven after which Eden was patterned!

The point is everything God created is orderly. He placed the family at the heart of what makes human life run properly. He is a loving Father, a Brother, and a Friend to those who know Him. He is warm and tender, caring for His own like a loving mother or an affectionate nurse. He knows how to assemble infants in a mother's womb. (Psalm 139:13) He made the universe, the Old Creation, into a comfortable and fitting habitation for His children. Those who are destined to inhabit the new house, His New Creation, will be those who get the moral lessons down right during this short life of trouble on earth.[2] For those who have not refused His grace and mercy, inheriting the universe is to be a grand, on-going, forever adventure. Nowhere does the Bible call this sort of thing myth, nor is Jesus an imaginary man who might have once lived in Israel. The future God has prepared for us is designed for men and women made whole and new by His work in their lives. Among all the multitudes around us almost everyone has chosen unreality to live in, yet in His love for all men, everyone is free to come to the cross of Jesus and enter into God's Greater House and into eternal life.

> 'The Spirit and the Bride say, 'Come.'
> And let him who hears say, 'Come.'
> And let him who is thirsty come,
> let him who desires
> take the water of life without price."
> (Revelation 22:17)

## Notes To Chapter One

1. Special thanks to Ray C. Stedman for this beautiful illustration.

2. "In a great house (the universe) there are not only vessels of gold and silver (people), but also of wood and of earthenware, and some for noble use and some for ignoble. If anyone purifies himself from what is ignoble (base), then he will be a vessel for noble use, consecrated and useful to the master of the house, ready for any good work." (II Timothy 2:20-21)

# Chapter Two

## Science and Revelation

### Part I: Science

Before launching into theological issues, I felt it worthwhile to discuss briefly the various and contrasting ways that knowledge and information reach the mind of man. Broadly I have divided these two categories into "science" and "revelation". The former is truth discovered by man's own search of the natural world, by experimentation, and by discovery that does not directly depend upon an outside intelligence supplying information. The latter is, I believe, knowledge that God gives to man by His own volition. Much of this knowledge is not accessible to man in any other way, except by revelation. Scientific "truth" cannot be, in the end, inconsistent with revelation, but gaps between these two areas of experience can and do occur. The existence of tension indicates only that we know now in part and not in full in either area. Because many readers will not necessarily have a scientific background, I felt it worthwhile to give my own view, based on experience, of how science operates. Because I am a Christian, I believe also in revelation from God, and I value this source of information more highly than that truth discoverable by science. By revelation I am, of course, talking about genuine revelation from God, not about visions or drug-induced experiences that could bring false information from a source which is not in the God of the Bible.

### Where Science Came From

Several years ago, a colleague and I were invited to present a brief discussion of the issues involved in the creation/evolution controversy raging in School Boards of Education and the Courtrooms of several states. My friend Kevin Wirth of Students for Origins Research constantly researches the published literature in this field of strongly polarized opinions and longstanding disputation. He knows much more than I do on the subject, so I limited my remarks to some general comments on science and revelation. However, in putting my notes together for our talk, I realized that not all the sciences enjoy equal credibility nor are able to make predictions with the same degree of success. Next I began to realize how much we owe to Biblical revelation for our background knowledge of the universe, though many today will not acknowledge any debt to revelation from God as a source of legitimate knowledge and understanding.

Though it is true that modern scientific thought in the West springs from Greek Philosophy, there is a debt to ancient Egypt, to Arabic and to Hebrew culture as well, for what we believe about ourselves and our universe in western civilization. Today there are some scientists who would draw on Hinduism or other Eastern schools of religious and philosophical thought to help us better understand modern physics and the nature of the cosmos. What many of our contemporaries seem to have forgotten is that the scientific discoveries of today rest on the foundations of previous generations, and many "founding fathers" of science in western science were either Jewish or dedicated followers of Jesus Christ and students of the Bible. (This is especially true in my profession which is physics and less true in psychology, for instance). Science as we know it in the West is a product of Western Civilization and the roots of this heritage are unmistakably Biblical in origin.

## Sifting Data to Find Order and Meaning

Our English word "science" is derived from the Latin *scire* which means "to know" or *sciens*, "having knowledge". Scientific endeavors in all disciplines are attempts to classify evidence so as to be able to make meaningful predictions. First, the data on the subject of interest is gathered and sorted. If the quantity and quality of the data is sufficient, then very frequently one can find patterns, trends, or order in what was at first a mere list of numbers or a collection of measurements.

If order can be found in the data, then laws can be formulated. These laws can then be extrapolated (inferred) beyond the bounds of the data at hand, and tested for accuracy, and examined for exceptions. Most of the sciences work better when the body of evidence is large and the data can be sorted statistically. One-of-a kind events like a reported resurrection from the dead, or five flying saucer incidents are difficult to study by any scientific discipline.[1] So also are happenings that are not reproducible consistently or which cannot be checked out by independent investigators in another lab.

The above principles severely limit what science can and can not tell us about our world.[2] On many subjects science must remain silent or carefully qualify those tentative theories that are put forth with little qualifying evidence to support them. There is no harm in putting forth hypotheses to be tested; indeed, many new useful theories come about this way. But most hypotheses are found-upon a little further investigation-to not fit the evidence and can be discarded in favor of those that do. It should be obvious, but to many it is not, that all science rests upon religious or philosophical presuppositions of one kind or other. Scientists who are secular humanists, atheists, eastern mystics, or Biblical creationists each bring their religious and philosophical presuppositions with them when they work in the lab. It simply is not possible to have science in a vacuum. One always begins with one set of assumptions (or another) about the nature of reality.

Every scientific model starts with a set of "initial conditions" and behind those, a whole set of notions about underlying or previous states of reality. As long as a scientist conducts his work by the rules of science, he or she should never be shut out of the lab, or the classroom, because of a belief system that differs from that of the majority, or one that is at the moment not the most popular point of view. One prominent professor of physics and astronomy has rightly said, "Science is in the business of discovering what the laws of physics are, not why those laws were passed. The latter is the realm of theology."

## The Rule of Simplicity in Science

All the scientists I know of assume the universe is orderly, as opposed to nonsensical, and wherever possible the simplest hypothesis is chosen because it is more "elegant" and saves extra work. This important principle is known as Occam's Razor. Thus "grand unified theories" are much to be preferred to complex theories because of this built-in assumption we make about the nature of things. In the West, scientific endeavor is based on a meaningful universe where the laws of physics change slowly, or not at all, and where data can be systematized so as to make useful predictions.

## Science Must be Meaningful to be True

The whole notion that the universe is orderly, that physical laws do not change, and that the explanations of things ought to be simple rather than convoluted (complicated), complements our inborn aesthetic senses, such as love of beauty and our personal hopes that a purpose for our own existence might be found. No one really wants to learn conclusively that life makes no sense at all for any reason. Such a conclusion leads to existential despair, a sense of futility, fatalism, and a feeling that human endeavors of all kinds are worthless and of no value. Among scientists there is always some hope and often a sense of great excitement and personal reward when meaningful patterns are found and previously uncorrelated information is found to "fit the curve."

## The Role of Intuition and Initial Assumptions

Intuition as well as observation is valued in science. Great discoveries have come from following hunches or making a "bold leap of faith", as Einstein did at the beginning of our century when he reached beyond the built-up body of evidence to date by means of intuition. Scientific knowledge is a body of truth arrived at by extending the ranges of the five human senses. Such things as microscopes, telescopes, infrared cameras, radio receivers and seismic monitors do just that: they allow us to peer into very real portions of our universe we would otherwise know nothing about.

All scientific theories are built upon assumptions, as I mentioned above. These foundational premises ought to be re-examined often since many times in the past tall palaces of speculation have been built on questionable and unproven assumptions. Scientific "advances" are built on the pioneering work of those who have gone before. If the pioneers made mistakes, or were short-sighted, their errors can easily be perpetuated for several generations. After a generation or two, the current scientific workers in a given field usually have "forgotten" or not taken the trouble to find out what assumptions went into the original work. Some have not bothered to ask

24

whether or not the data base has changed or checked to see if the original assumptions are now suspect or erroneous. And yesterday's speculation becomes today's scientific dogma in many instances. No tenured professor drawing a comfortable salary and enjoying a maturing successful career is likely to be objective, or even very rational, if a newcomer to his field questions the evidence and finds the professor's whole theory must be thrown out the window in the light of new evidence. Yet this process happens all the time, silently, as one generation fades away, new "authorities" come to power, and better theories take the place of the "primitive" notions that were held as absolutes in the previous generation.

## Theoretical and Experimental Branches in Science

Some scientific discoveries have been the result of purely theoretical studies conducted by mathematicians. Experimental testing of such theories has in many instances led to valid new knowledge. In fact, science divides broadly into classes of theoreticians and groups of experimentalists (who need each other if only to stay honest and realistic in what they undertake). Theoretical studies, such as mathematical models of the universe, allow for many more dimensions and variables than may *actually* exist in the known universe, yet some scientific discoveries have been made purely because some mathematician suggested that something in one of his equations might help us understand a previously poorly understood area of science. Upon investigation, the suggested phenomenon (a new atomic particle, for example) has often been found to exist. Researchers, that is, experimentalists, frequently find discrepencies in their measurements that lead to new or better theories. Or, by accident, they may stumble on to some previously unexplained phenomenon. When this happens, they call in the theoreticians, who must go back and do more homework. Though each of these two groups claims to have the superior point of view, it is obvious that a synergy between them exists, and their interaction with each other from differing points of view is most valuable to us all.

# The Reliability of the Sciences

One of the most interesting things about science in the western world today can be discovered with the help of a little diagram. If one makes a list of several scientific disciplines - grouped by category somewhat as shown in the following table - it will be immediately obvious that science knows *most* about the physical, inanimate world, and *least* about the world of spirit:

| The Relevance of Science in Various Areas of Human Experience | | | |
|---|---|---|---|
| **Physical World** | | **Emotional World** | **Inner Man** |
| External to Man | Life Processes | Mind, Emotions, Will (the soul) | Realm of the spirit (the 'heart') |
| Physics Astronomy Geology Geography Meteorology Chemistry Archaeology Mathematics | Biology Botany Zoology Medicine Organic Chemistry Anthropology | Psychology Sociology Criminology Social Sciences Behavorial Sciences | ESP Parapsychology Astrology Metaphysics Occult Sciences |

------------------------------------------------>
**decreasing scientific reliability**

The most important conclusion to be drawn from such a table (to my way of thinking) is that those sciences that attempt to deal with the spiritual realm (that zone closest to the innermost recesses of man) have not proven very successful in spite of valiant attempts to make them so. They are frequently discredited by conservative scientists or considered "pseudosciences." But no one argues very strenuously with the findings of the astronomer or the geologist in comparison. There we rest on empty space or solid ground respectively! This way of looking at things seems to be more comfortable to us! When measuring and studying the physical, material world, our instruments are more reliable; our results are more easily reproducible; and our findings tend to better stand the test of time.

It is easy for us to put our faith in what can be seen and touched, measured and felt - and not so easy to look into the invisible, to trust a God we can't see, or even to think about who we are as living beings having emotions and spirits as well as bodies. It is of course easier to love another human being we can see and touch - than to love God Who is invisible.

Psychology lies in between the outer world of matter and the innermost world of spirit, but since it is difficult to systematize dreams and build up theories of neuroses and psychoses in the rational language scientists accept, the psychologist is as much an artisan as a scientist. If a psychologist is "good," it is because he succeeds in healing someone whether or not anyone knows how he does it.

In spite of his atheistic beliefs, Sigmund Freud and others who followed after his example have at least tried to systematize their findings into a collective body of knowledge so as to build up a body of empirical, and therefore useful, knowledge. Much of modern psychology is, therefore, a legitimate science. A good many contemporary psychologists have, unfortunately, discarded revelation. In doing so they are attempting to reduce man to a merely physical entity. Perhaps I should add that almost all scientists are pragmatic at times and sometimes even eager to seize upon a formula or discovery and begin to apply it according to the unwritten law "If it works use it. Never mind why. We'll figure that out later."

We are only just beginning to understand the workings of the brain (let alone the mind), and attempts to explain man's innermost workings on the basis of physics, chemistry, and electricity make us seem like extensions of the material world. Few of us want to have our souls stolen from us so that we are no longer thought of living, caring, vulnerable beings for whom beauty, art, music, culture, lov are more important than all things scientific. We even like it better, or ought to I believe, when we think that we live in a universe full of mystery, with room for endless discoveries as well as endless delights. In the long run, science really only helps us in the physical world and in the biological world, that is in the world of externals, the things of the material half of creation. We get no real help from science on metaphysical matters. Even concerning the material world, science knows very little indeed!

## Science Is Limited (Mostly) to Physical Realms

All our public preoccupation with science in the western world ignores a fundamental statement of the Bible, namely that the physical world is a world of shadows and that the real and permanent world is invisible. This is stated in several ways, especially in the New Testament. For example, Paul writes, "**We look not to the things that are seen, for the things that are seen are transient, but the things that are unseen are eternal,**" (I Corinthians 4:17) and, "**...the form** (schema) **of this world** (kosmos) **is passing away.**" (I Corinthians 7:31)

I mentioned the greater importance of the spiritual compared to the material creation in Chapter One. As I said, most of us grow up with the notion that the physical world around us is solid, substantial, and enduring, whereas the spiritual world is the world of ghosts and shadows - but, in fact, the Bible declares that the opposite is true! To help turn my own thinking around towards the right way of looking at things, I have benefited much from books such as those of C. S. Lewis. For example, his *The Great Divorce* gives a soundly biblical idea of what heaven might be like as shown to us in Scripture. Life as we know it on earth pales into insignificance compared to what heaven will be like!

It is the excessive preoccupation of science in the West with tangible reality, as opposed to the spiritual source, that I believe has helped dry up the soul of modern man. We have also led ourselves into unmanageable contemporary moral issues whose resolution can be found only in the spiritual. I for one do not wish history rolled back to an antiscientific era, nor am I willing to relinquish the many conveniences modern science has brought us to improve both human effectiveness and enjoyment of life. In fact, the Bible says that God has given us all things freely to enjoy. (I Timothy 6:17) So it is unbiblical asceticism to deny the physical and strive only for the spiritual. But science cannot do everything, and it should really not surprise us if the old order falls and a new age enters. It has happened before in history. A "merely physical" model of the universe has always proven inadequate. Sooner or later what was once thought "solid science" is seen to be fallacious and incomplete if God is left out of the equations. Fortunately He has found ways of breaking into our simplistic fantasies about ourselves just when we need Him most.

## Part 2: Revelation

The usual Greek word translated "revelation" is *apokalupsis*, which means "unveiling". A related New Testament word is "mystery" (*musterion*) - and there are about seven or so "mysteries" categorized in the New Testament. A mystery in the Bible is a body of truth that was hidden in a previous age but later can be understood clearly if one takes the time to examine it. Just to cite one example, the relationship between man and wife in marriage is likened to the relationship between Christ the Bridegroom and His Bride the true church; this is said to be **"a great mystery".** I am sure none of us have a problem seeing the depths of this this subject. "Mysteries" in the Bible are complex bodies of truth capable of being understood, though multifaceted and even full of paradoxes. The concept of "mysteries" in the Bible suggests clearly that our universe is indeed so complex it cannot be fully understood by any one discipline or field of knowledge. The biblical notion of "revelation" assumes that man is largely "in the dark" about himself and his universe (as a result of the fall) and needs much outside help and enlightenment.

### The Self-Revealing God who Created Us

Without revelation we might know nothing about the nature of God Himself, for He could keep Himself entirely hidden from us if He had chosen to do so. As it is, the Bible contains an enormous body of knowledge about the nature, character, attributes, and activities of God. Furthermore the Biblical record, unfolding as it does over the space of hundreds of years, tells us that the revelation of God is progressive in history. At one point in history God Himself entered our domain as the man Christ Jesus; a hundred years later the New Testament was completed; at a future day Christ will return in visible splendor and glory, and so on. It was Jesus who said that **"all the hidden things of darkness will be brought to light,"** and **"there is nothing secret that will not be made known."**

Whether we like it or not, the God who created us intends eventually to bring the full truth about everything and everyone out into clear public view, letting the chips fall where they may. The false will give way to the full truth, and our incomplete understanding will be made complete. (I Corinthians 13) I mentioned that divine revelation has been progressive in time. This also means that God's revelation of Himself during the course of history has gradually increased. The writing down of scripture, for instance, was not the work of one man nor was it all given to inspired authors at one time. The presentation of Jesus as God's "final Word" to mankind came "late" in history, near the end of an age.

## The Amazing Reliability and Accuracy of Scripture

The whole idea that the Bible is an inspired book does not mean that the writers were merely visionary or full of unusual enthusiasm about the subject matter. Scripture says of itself that it is the very Word of God written by specially-chosen men consecrated for the purpose of setting down a totally self-consistent body of revealed truth that would withstand all possible scrutiny and prove reliable in detail for all time. The Hebrew Kabbalists love to study even the numerical and mathematical models of the Tanach (the Old Testament) because a numerical value can be assigned to every letter of the Hebrew alphabet, and this allows one to do wondrous things with the text itself that are mathematical, and geometrical. It is quite true that the Bible is full of numerical, geometrical, mathematical and symbolic truth in addition to the clear meaning of the text. Unfortunately some ignore the main purpose and intent of the writers of the Bible which is primarily directed towards improving the human moral and spiritual condition on the planet. However, when the Bible mentions matters of science, it is always consistent with the universe as we know it today from modern science. For example, the fact that the earth is a sphere suspended in space is mentioned in the oldest book of the Hebrew Bible, the Book of Job. Perhaps one should say that the Bible is not preoccupied with science, but as the narrative unfolds, we are given a great deal of peripheral evidence that is scientifically valid if we will but take the time and trouble to search it out.

Biblical revelation concentrates on what is wrong with the human race, and I suppose many of us like to avoid the doctor's office until the pain and confusion become too great to bear. Then, "if all else fails, follow directions," as a pastor friend, Bob Smith, used to say. The Old Testament is especially poetic in its imagery, but that does not mean that profound scientific truth is not tucked away there for us to find. The Bible has a lot to say about the complex nature of time, for instance, as divided into "times" and "seasons" as well as hours, minutes, and seconds as science has chosen to do. The Bible *is* an Oriental book so this may be the reason we yearn for more "knowledge from the Eastern side of things" showing up on every side in the West today. Our insistence that all truth be rational, analytic, dissectable and ordered, and verifiable in the laboratory misses the Biblical view to a large degree. Nor does the Bible teach on a given subject all in one chapter or book. Instead, bits and pieces are given here and there, and the only way to grasp a subject clearly is to read (and re-read) the Bible from Genesis to Revelation day after day and year after year. King Solomon said in Proverbs, **"It is the glory of God to conceal things, but it is the glory of kings to seek them out".** (Proverbs 25:2)

However, the Bible is meant to be understood even by children, for the truth about very difficult subjects is approached by scripture in terms anyone can understand. This fact about Biblical truth, therefore, *does* support the scientist's wish to have a universe that can be understood in simple terms rather than complex ones - whenever possible. The Bible describes the universe as a *uni*verse, a unified whole entity; therefore, if we understand things on a small and local scale, we can likewise understand the large scale and the very small and the invisible and the remotest corners of space. We can know ourselves and we can know our Creator!

## Man's Central Place in God's Creation

Central to Biblical revelation is the notion that man, made in the image and likeness of God, was placed into the center of creation[3] to understand it and to rule it. For example, when Adam was given the task of naming the animals God brought before him, it was to be a training session to help Adam understand what God had made, so that he, Adam, could govern wisely as the representative and steward of God over creation.

Biblical revelation tells us that societies prosper and increase in knowledge and understanding because of the grace of God - not because of human merit or effort. Likewise, so-called primitive societies are examples of groups of people who have ignored God's

31

grace and thus fallen into degradation and ignorance. The Bible also teaches that "progress" as understood by modern man is largely an illusion! Sometimes God blesses a nation to honor previous generations of godly men and women. Sometimes He raises up a great nation such as Ancient Egypt as an object lesson, or to demonstrate His displeasure at the behavior of His chosen people. The fall of man was downward into total depravity; civilization is the thin veneer that hides the savage beast. "Science" is not building a better world for our children, but God continues to bestow many good gifts on those who believe in Him and much "common grace" upon those who do not.

## Truth, Like a Diamond, Has Many Facets

Because the universe is so complex (as seen by the Bible), it is often necessary to consider "both/and" rather than "either/or". Thus, it should be no surprise that truth is found on both sides of a marriage dispute, in a case before a court of law, or in a creation/evolution argument. In the familiar picture of the elephant as described by three blind men, one man described the elephant as being like a rope; the second, like the trunk of a tree; the third, like a wall. We know that all three men were partially correct, but that the elephant was much more than even the sum total of the three pieces of evidence presented by the blind "scientists".

## The Grand Deceptions Man Lives By

Not only is our data base incomplete, but the actual world we live in is full of active evil, of erroneous reasonings, of grand deceptions. And the best of us, as persons, is highly flawed. The Apostle Paul writes of human ignorance and the ease with which most of mankind is deceived in every generation - beginning with our leaders: scientists, political leaders, and barons of industry:

**"Where is the wise man? Where is the scribe? Where is the debater of this age? Has not God made foolish the wisdom of the world? For since, in the wisdom of God, the world did not know God through wisdom, it pleased God through the folly of what we preach to save those who believe. For Jews demands signs and Greeks seek wisdom, but we preach Christ crucified, a stumbling**

block to Jews and folly to Gentiles, but to those who are called, both Jews and Greeks, Christ the power of God and the wisdom of God. For the foolishness of God is wiser than men, and the weakness of God is stronger than men...And I was with you in weakness and in much fear and trembling, and my speech and my message were not in plausible words of wisdom, but in demonstration of the Spirit and power, that your faith might not rest in the wisdom of men but in the power of God.

"Yet among the mature we do impart wisdom, although it is not a wisdom of this age or of the rulers (leaders) of this age, who are doomed to pass away. But we impart a secret and hidden wisdom of God, which God decreed before the ages for our glorification. None of the rulers of this age understood this; for if they had they would not have crucified the Lord of glory. But as it is written, 'What no eye has seen, nor ear heard, nor the heart of man conceived, what God has prepared for those who love him,' God has revealed to us through the Spirit. For the Spirit searches everything, even the depths of God. For what person knows a man's thoughts except the spirit of the man which is in him? So also no one comprehends the thoughts of God except the Spirit of God. Now we have received not the spirit of the world, but the Spirit which is from God, that we might understand the gifts bestowed on us by God. And we impart this in words not taught by human wisdom but taught by the Spirit, interpreting spiritual truths to those who possess the Spirit. The natural man (man in his natural condition before spiritual rebirth) does not receive the gifts of the Spirit of God, for they are folly to him, and he is not able to understand them for they are spiritually discerned."
(II Corinthians 1:20-25; 2:3-14)

This remarkable passage of God's Word helps us to understand man's arrogance and self-deception especially in areas concerned with knowledge, power, and influence. It is not, however, knowledge alone that gets us in trouble, it's the application thereof. It is wisdom we lack, and the experienced insight we need to run our lives, and to understand things properly, can only come to us from God - and then only when we are willing to be enlightened and instructed by Him. Science itself is morally neutral, and one can readily say from the Bible that God is not opposed to science or the increase of knowledge. Quite the opposite. He is the very God of all knowledge and understanding!

Our estrangement from God and continuing rebellion against His guidance in our affairs inevitably means things will go towards the worst possible outcomes, nuclear winter included. In fact, rejection of revealed truth, common in our land today, causes God to give us over to what is called in the Bible *the* lie. This is a deep belief in one's own self-sufficiency, in the truth of prevailing science and philosophy (even when they are out of tune with Scripture), and the notion that man is his own god and master of his own destiny. Hearing truth and rejecting it leaves us worse off than remaining ignorant of truth, though neither state of affairs excuses man from accountability to his Creator.

A personal commitment to Jesus as Lord is, of course, what we need. It is that step of faith and commitment that restores us to a relationship with the living God so that we can perceive reality - the way things really are, in our hearts and minds and lives. Thereafter we can begin to participate in what God is now doing in the universe, which is to build a *new creation* and a new race while the *old creation* fades away and comes to an end.

What Jesus said about truth (as recorded in Scripture) is that **He Himself is the truth!** This is an amazing statement - that is, that truth is a person. The Apostle Paul adds, **"In Christ are hidden *all* of the treasures of wisdom and knowledge."** (Colossians 2:3) Were it not for scripture, men would think all truth was relative; and we would be unaware of error, false teaching, and the widespread, deceptive nature of false philosophies everywhere on our planet. We all tend to fall into the trap of believing that truth is contained only in doctrinal, philosophical, or scientific statements about the nature of reality. That absolute truth is contained in a Person assigns a fundamentally different meaning to the Greek word *logos* than was found in Greek philosophy prior to the writing of the New Testament. *Logos* is one of the titles assigned to Jesus as the Creator in the well-known preamble to John's gospel:

**"In the beginning was the Word (*logos*) and the Word was with God and the Word was God. He was in the beginning with God; all things were made through him and without him was not anything made that was made. In him was life, and the life was the light of men."** (John 1:1-4)

## Truth from Revelation not Accessible to Science

The Bible tells us we are surrounded by myriads of angelic beings, a statement that no reputable scientist has put forth nor attempted to prove (or disprove) scientifically. Were it not for scripture, we would think of the spiritual realm as ghostlike or "imaginary" rather than permanent, substantial, and enduring. Scripture tells us that man is not just a physical being, but also a spiritual being, and that man is, above all else, a spirit. Though human evil is everywhere evident to anyone who is honest about himself and the world, this fact is constantly denied in the halls of science, in schools, and in political campaigns - denied almost everywhere by men, but not by the Bible. The Bible confirms that the universe is an orderly, composite structure but goes further by making clear our Creator's goals and purposes in both creation and redemption. Scripture tells us clearly where history is heading and how things will end. Secular science and secular history see none of this accurately. Rather than man's being incidental to creation, the Bible places him squarely in the center as the highest of created beings, intended by God to exercise dominion over the earth, and later over the new creation God is now building. It is the Bible that explains us to ourselves and provides the remedy for human evil so that healing and restoration can begin within us. Where else but in Holy Writ can we learn about the causes of war, the reasons for upset ecologies, the purposes of government, and the "Four Last Things" (Death, Judgment, Heaven and Hell) that no man can avoid discovering in experience in the end?

Finally, it should be clear that I hold that truth, properly perceived as revelation from God, is absolute truth and can be depended upon fully to guide a person in this life and in the next. Scientific truth may or may not hold up under the tests of time and comparison with absolute truth. Unless it does, scientific truth has limited value and limited scope.

## Christ the Wisdom of God

It is interesting that Jesus Christ in addition to being called "the truth" is also said to be the "Wisdom of God" in the New Testament. Wisdom in the Old Testament book of Proverbs is personified by a woman. When one speaks of wholeness for "mankind" as an ideal or a goal, we need to remind ourselves that fifty percent of life is represented by the feminine point of view and that the creation of man in the image and likeness of God was the creation not of Adam, the man, but of Adam/Eve together. The Biblical revelation that both God and His creation can only be understood by combining a masculine and a feminine point of view, of course, is an idea foreign to most fields of western science. Imbalance in science is a result of leaving out such alternate (and complementary) ways of looking at reality. Of all the many problems the evolutionist faces, one of the greatest is how to explain the existence of sexuality in creation at all. It is only when we turn to the Bible that we gain light on this subject.

The often quoted saying, **"now we see through a glass dimly, but then face to face,"** comes from the great chapter on love found in First Corinthians 13 and is a reminder that of all that God has told us about Himself, He assures us that love is of central importance. We are loved by our Creator and He has paid the highest possible price by giving His Son, our Lord Jesus to make it possible for us to inhabit His universe forever. Science does not tell us this, but revelation does. To make science a religion in place of the reverence, worship, and devotion to a God of Love, who is at the same time Truth itself, is to be deceived and enslaved by a terrible idolatry indeed.

In some ways modern science may be compared to ancient gnosticism. The Gnostics specialized in *gnosis* or knowledge, but the Apostle Paul in Colossians speaks of *epignosis*, or full knowledge. This is what Jesus Christ brings to all who seek Him and place their trust in Him. This is how we enter into the full dimensional understanding He intends for us. As scripture says, **"God desires that all men should come to a knowledge of the truth."** (I Timothy 2:4)

❒

# Notes to Chapter Two

1. A great service to science is provided by Mr. William Corliss who collects abstracts and reputable scientific accounts of the rare and the unusual. In addition to a number of volumes that Mr. Corliss has published by topic, he also stocks rare and difficult-to-obtain scientific books and publications (some out of print) for sale to his interested researchers. For a sample of his regular newsletter, *Science Frontiers*, and a price list, write *The Sourcebook Project*; P. O. Box 107; Glen Arm, MD 21057.

The subject of Unidentified Flying Objects (UFO's) has been of interest not only to amateurs and pseudoscientists, but to outstanding, conservative and careful scientists such as my friend Dr. Richard F. Haines. His latest book on this subject, *Melbourne Episode: Case Study of a Missing Pilot* (available from the author at 325 Langton Avenue; Los Altos, CA. 94022), is highly recommended. I personally accept the evidence that UFO's are real, and I suspect they are angelic phenomena. These may be divided into two classes: angelic activity by fallen angels under Satan and benevolent angels devoted to the service of God. It is also possible that they are intelligent beings from other planets; however, the Bible says nothing about such forms of life. Man is the highest of God's creations, and earth is Center-Stage for God's work in redemption and the consummation of history. Since many UFO phenomena are so difficult to explain by the existing laws of physics, they may well be intrusions into our physical world from the invisible world of the spirit which surrounds us. Angelic visitations from the spiritual world into the physical are common in the Bible. I do not think Ezekiel's "wheels" are necessarily spacecraft, however.

2. George S. Hendry in his *Theology of Nature* (Westminster Press; Philadelphia, 1980) presents an excellent discussion of the relation between God, man, and nature. He notes rightly that science and religion are not the only disciplines available to us to study the world in which we live.

Paul Davies, author of *God and the New Physics* (Simon and Schuster; New York, 1983), is a secular physicist who believes that science offers a clearer perception of reality than religion. His book contains an excellent bibliography.

3. A recent good book on the subject is *Origin Science* by Norman L. Geisler and J. Kirby Anderson (Baker Book House; 1987). Thousands of Christians who are also competent and capable scientists actively participate in such organizations as the Creation Research Society, The Bible-Science Association, The Institute of Creation Research, Students for Origins Research, the Creation-Science Research Center, and The American Scientific Affiliation. Many of the journals, publications, and newsletters of these groups are erudite and challenging, both scientifically and spiritually, and will stand the test of time. Mankind as a whole is at enmity with God, and it is no surprise, therefore that many "creationists" today are ridiculed or blackballed by their secular colleagues. Knowing God is, however, a life long process and the most knowledgeable Christian knows really very little, and less as he grows wiser! Creationists are not infallible, but God is!

A book representative of the vehement attacks on creationists these days is a collection of essays edited by Laurie Godrey entitled *Scientists Confront Creationism* (W. W. Norton and Company; New York, 1983).

# Chapter Three

## Is the Universe Running Down?

### A Universe Decaying and Fading Away

Actually, Biblical revelation adds to rather than detracts from our knowledge of the physical universe, provided one takes the trouble to search out hidden truths of scripture that reflect that part of reality that is accessible to scientific study. In this chapter, I want to discuss the question, "Is the Universe Running Down?" The issue is well-known to the physicist and very important to our understanding of how things "work" in the physical universe as we know it today.

The issue under discussion is known as the "Second Law of Thermodynamics." It states simply that as time flows forward, energy in the universe is becoming less and less available. Also, orderly systems tend - on their own - to become disorderly and chaotic through the processes of decay and disintegration that are part of the very fabric of the material world. Strictly speaking, the Second Law applies to what is known as a closed system, but almost any discussion of the applications of the Second Law can be considered by drawing a sufficiently large boundary around the components being studied. For instance, the earth by itself is not a closed system as far as energy input is concerned, but the solar system is, for all practical purposes, since little outside energy reaches the earth except from the sun.

Trained as I have been in physics, math, and engineering, I have grown up along with many colleagues and friends trusting that the "Laws of Physics" are immutable and trustworthy. Indeed, this is the normal experience of the physical scientist or engineer who confidently makes many predictions based on theories that work perfectly every time. If the laws of physics are broken or violated, such things have escaped our notice thus far! I may be naive but have always supposed physics to be a more "exact" science than biology. At least it seems easier to me to quantify and measure the properties of the physical universe and to extrapolate beyond the normal limits of the five senses when dealing with non-living systems.

Recently, however, in the world of physics, there has been new evidence that some of the "constants" of physics - such as the velocity of light - may, in fact, not be constant after all, but may have changed over time. If this proves to be the case (these issues are discussed later on in detail), then our ideas about the history of our universe will be radically revised based on science, even apart from revelation.

The Laws of Thermodynamics are discoveries about our physical world. The First Law observes that as far as we know, energy/matter cannot be created or destroyed, but only converted from one form to another. The Second Law has to do with the flow of energy in the universe, which is always observed to be "downhill" from states of high availability to states of low availability. Observation and theory show that the universe is indeed running down, like a clockspring gradually uncoiling as energy is dissipated by the escapement mechanism. Ordinarily, this means that heat flows from high temperature "reservoirs" to low temperature "sinks." To reverse the process, one must put in additional external energy to "pump" heat from a low temperature reservoir into a warmer. This is how refrigerators and air-conditioners work - to give two very elementary examples.

In either case, as time goes by, a quantity known as the total *entropy* of the system increases. Entropy is the physicist's term, a measure of the state of energy unavailability of an energy-containing system. According to the Second Law, the physical universe will one day suffer a so-called "heat-death": all the high energy (hot) reservoirs will have poured their energy into the colder parts of the cosmos, and everything everywhere will be at a completely uniform temperature. In such a state it would be impossible to build any more heat engines that could do useful work, and the universe would "die" or grind to a halt. Although this discussion has centered on heat energy, the same rules apply for electrical energy, magnetic energy, gravitational energy, mechanical energy, and so forth.

## Order and Information Content

Another way of stating the Second Law has to do with order and complexity in the universe. Orderly systems of molecules represent low entropy systems, and with the passage of time the normal tendency of things is for such systems to become disorderly, chaotic, and randomized. To bring order out of chaos, we must put in outside energy and also programming information - such as instructions, blueprints, or genetic codes. The process of utilizing energy is always less than 100% efficient. Energy is wasted every time we build and operate an engine or a machine to do useful work.

In the physical world we never see any significant natural examples of energy flow from cold reservoirs to hot, and only rarely do we observe any small (and very temporary), slightly more orderly arrangements of molecules coming into existence as part of the process of statistical fluctuation. The physical universe as we know it is characterized by rust, increasing ruin, and decay. Order comes from chaos only if someone makes it happen. Of course, man, being an intelligent creature capable of organizing things, has built for himself all sorts of heat pumps and engines and complex structures, constructing orderly systems out of random parts. In doing so, however, energy is required and the total entropy of the universe always increases. The Second Law is not violated in any of these activities undertaken by man.

## Living Systems and Entropy

When we take living things into account, what appears at first glance to be an exception to the Second Law occurs. Living cells are observed to build orderly systems out of simple molecules assembled, one by one, according to inherited, self-contained genetic (RNA/DNA) instructions. Living systems do not really violate the Second Law because all living things burn energy as fuel. The built-in genetic codes that living cells carry provide the intelligent instructions for assembling orderly systems out of simple molecules.[1] Energy inputs from the physical universe are required to run biological engines, so all living things live at the expense of raising the total entropy of the physical universe. The intrinsic high complexity of the genetic codes implies a

Designing Intelligence at the beginning of the universe who brought into existence these original blueprints in the first place. The Second Law disallows these codes having come into existence by accident. Time plus chance always leads to chaos - not order - without the intervention of outside intelligence!

As far as I know, no biologist can tell us how a living cell differs from a dead one - except that the upbuilding, growth process ceases, as do repair, replacement of cells, and reproduction when death occur. At death the once-orderly cell decays back into the elementary molecules from which it was assembled. From Scripture we learn that life[2] is due to *spiritual* energy activating cells. For life to exist, some form of energy supplied by God apparently flows into our physical world from "outside the system" of classical thermodynamics.

## Natural and Spiritual Energy and Power

A hundred years or more ago, some early scientists tried to weigh the human soul by placing dying persons on delicately-balanced scales. No one was able to detect any weight loss at death due to the departure of the inner man. Beyond these primitive experiments to weigh the soul, I know of no experiments ever made that attempt to determine if measurable spiritual energy or power from living persons or living things flows into our physical universe. This is a most important question because in the Bible *life* and *light* are often placed in close relationship with one another; for example, in the Prologue of John's gospel: "**In him** (the Word) **was life, and the *life* was the *light* of men.**" (John 1:4) Although this is a statement about spiritual reality, as revealed in Scripture, there is little doubt that the great Light, the Person Jesus Christ, is behind all forms of ordinary light, and light in our physical world is always a form of energy. Spiritual energy is actually the source of all *other* forms of energy. Behind everything in our physical world of mere shadows, stands the permanent, enduring reality of the spiritual. The Bible speaks also of the "mighty power" God the Father exercised when He raised Jesus from the dead and brought him into heaven forty days later. Yet, I know of no meter or instrument or scientific recorder that could have been placed in the tomb of Jesus to "detect" and "quantify" that "mighty power." Yet, the power of God, who is the Source of all life and energy, is much more important than the flow of energy in kilowatts or joules or horsepower studied in the everyday world of physics.

In the moral sphere it is easy to see that the decay of civilization, family, and culture is held back and inhibited by true spirituality, by God's intervention in our world of death and decay. What this means is that available Christians are channels of "light" and restraining "power" against evil flooding into a "dark" world. In the spiritual realm of soul and spirit, new life from Jesus does very definitely affect human life and society. Such changes in human behavior are not illusions, but are certainly impossible to measure in terms familiar to even the sociologist or the psychologist. The subjective measuring rods of the latter two disciplines are valuable of course, but physics has to be content with the physically measurable, and that limits one's perception of reality to a considerable degree. Only the eyes of faith see things are they really are: **"By faith we understand that the world was created by the word of God, so that what is seen was made out of things which do not appear."** (Hebrews 11:3)

## The Spiritual and Physical Worlds are Connected

Further evidence of a connection between the spiritual life of man and the power flow from God into the universe can be found in Genesis. As created, the universe was pronounced "good" by God. This means a high degree of order originally existed. Another way of saying the same thing is that the universe had low initial entropy. However, creation proceeded in a series of steps as revealed in the Bible - not in a single creative act, which is what the "Big Bang" hypothesis suggests. In Chapter One, I emphasized that a created invisible order (the heavens) and a created visible, physical world were both brought into being at the same time. From the beginning, these two realms were interlinked one with the other in complex ways. Heaven is the permanent reality; earth, the derived world of shadows and illustrations. Man has access to both realms of creation through a relationship with God, and it is largely through man that the physical created order is linked to the invisible world of spirit.

Before the fall of man, it is quite likely that the energy reservoirs of the *physical world* were all initially filled when the universe was created. Perhaps God always kept them replenished, continually renewed, until the fall of man, because the physical and the spiritual were, until the fall, "close-coupled systems." It is clear, however, from the Genesis account that the *power flow of spiritual energy into nature decreased abruptly after the fall* - a permanent change occurred. No longer was nature to produce bountifully for man; extra work and toil would be required, and weeds and pests would interfere. Death entered through the man Adam, Paul writes to the Romans. The entrance into our world of death, dying, decaying, destroying processes affected not only man but also nature as well. I believe this thesis can be further established by a well known passage in Romans 8:19-23:

**"...the creation waits with eager longing for the revealing of the sons of God; for the creation was subject to futility, not of its own will but by the will of him who subjected it in hope; because the creation itself will be set free from its bondage to decay and obtain the glorious liberty of the children of God. We know that the whole creation has been groaning in travail** (labor-pangs) **together until now, and not only the creation, but we ourselves, who have the first fruits of the Spirit, groan inwardly as we wait for adoption as sons, the redemption of our bodies."**

This scripture tells us that the old universe, what the Bible calls the "old creation," is fading away, dying and decaying. The Bible also indicates that God's renewing work in our present lives involves the soul and the spirit, but the physical body is not made new until the resurrection. Although the Holy Spirit gives life to our mortal flesh also, healing and sustaining our physical bodies as we walk with God, the dying of the outer man, the "old Adam" is only delayed, not reversed, when we become believers, **"So you also must consider yourselves dead to sin and alive to God in Christ Jesus."** (Romans 6:11) God may not be actively supplying any new energy into the physical universe, the Old Creation, but rather allowing it to run down according to the Second Law. Thus we say that it is "natural" for the universe to be running down, that is the way things are.

This raises the difficult question of miracles. Many miracles in the Bible are simply examples of God's speeding up natural processes (turning water into wine, for instance). As Lord of Creation, Jesus has the power to intervene in the old order of things; but the question is, does He do so by putting new outside energy into the system, or does he simply alter the arrangement of the old order in a small way whenever he performs a genuine miracle? Is His creative energy input directed mainly towards the building of a "New Creation" instead? To consider such

44

issues adequately, one needs to explore a discussion such as that given to us by C. S. Lewis in his book *Miracles*. Lewis offers many helpful comments on these difficult matters.

Even after much reading though, I do not find many writers, secular or religious, who discuss the interlinking of spiritual and material realms of creation and the existence of any "laws" governing the two realms taken together instead of separately. There are spiritual laws just as there are physical laws, but I do not know how to "measure" the mighty power God manifests in nature or in human affairs. I really do not know if some "miracles" are violations of the laws of thermodynamics as we know them, or whether God only directs the present universe (the Old Creation) by working within the bounds of the existing laws of physics. To "override" gravity by spiritual means, it is not necessarily to "violate" the laws of physics, yet it may well be that many genuine miracles do represent God's intervention in the old order of things for His own specific purposes.

## Hidden Variables and the "Laws" of Chance

As a digression for the interest of more technical-minded readers, one of the great questions raised in the branch of modern physics known as Quantum Mechanics is whether or not there are "hidden variables" in physics operating in the "noise level" below the range of sensitivity of our instruments. The existence of such laws would allow our universe to be influenced without our seeing how it was done. At the sub-atomic level, Quantum Mechanics has shown that statistical fluctuations apparently govern all physical phenomena. For this reason some men continue to believe that the universe came into being solely out of the operation of time-plus-chance. Electrons behave at times like particles and at other times like waves. The presence of an observer disturbs the status quo so that one cannot measure anything in the universe without at the same time disturbing its state of being. If one chooses to measure one parameter of physics with precision, another related parameter automatically becomes proportionally fuzzy and indeterminate!

The meaning of the so-called "Uncertainty Principle" is quantitatively expressed by such equations as

$$\Delta p \, \Delta x \geq h$$

where $\Delta x$ is the uncertainty in the position of a particle in the x direction and $\Delta p$ is the uncertainty in momentum, p, (mass times velocity). Other pairs of variables besides p and x may also be used. The two uncertainties are related to h, Planck's constant. As will become clear in Chapter Seven, if the velocity of light has been decreasing since the universe began, then Planck's constant has been *increasing* in inverse proportion to c. (In fact, the product hc seems to be a true constant.)

Any decrease in the velocity of light since creation means that h was a smaller number in the past than it is now. If h were smaller, a number of atomic proccesses would be affected. Alpha particle emission from heavy nuclei, for example, depends on "tunnelling" of the alpha particle under the curve of nuclear binding potential. Such tunnelling depends on "borrowing" extra energy $\Delta E$ for a short time $\Delta t$, however subject to $\Delta E \, \Delta t \geq h$. Thus alpha particle decay of atoms would have been more probable (and therefore more frequent) in the past.

If h were zero, we could also say the universe was perfectly "determined" and more "certain" or predictable than it is now. We could then make perfectly precise measurements. In fact, with h not equal to zero, the measurement process always disturbs the system under study and makes precision determinations impossible. Thus, if h were a smaller number in the past than it is today, due to c-decay, the universe would have been more orderly in the past. We might say that this was due to "perfect coupling" between the spiritual world, which is the source and the physical world where we make measurements.

## God Controls Everything

Actually Scripture has two things to say in relation to such matters as hidden variables and the laws of chance. In Proverbs, **"The lot is cast into the lap, but the decision is wholly from the Lord."** (Proverbs 16:33) This and related passages show that the physical universe is *not* actually governed by random, statistical events. Rather, an Unseen Hand - working behind the scenes - is directing the affairs of men, nations, nature. In the Sermon on the Mount Jesus assures the multitude that the hairs of our heads are numbered, that God clothes the fields with grass and feeds the birds of the air. Furthermore, although we search diligently to

46

find out *how* God does things, we cannot know the answers to that question:

"O the depth of the riches and wisdom and knowledge of God! How unsearchable are his judgments and inscutable his ways! 'For who has known the mind of the Lord, or who has been his counsellor?' 'Or who has given a gift to him that he might be repaid?' For from him and through him and to him are all things. To him be glory forever. Amen." (Romans 11: 33-36)

The 102nd Psalm also gives us further help in understanding the present state of the created order of things,

"Of old thou didst lay the foundations of the earth, and the heavens are the work of thy hands. They will perish, but thou dost endure; they will all wear out like a garment. Thou changest them like raiment, and they pass away; but thou art the same, and thy years have no end." ( Psalm 102:25-27)

What is clear at this point is that both the Bible and the Second Law describe the known universe (that is, the "old creation") as decaying, "wearing out," and coming to an end. If God is intervening from time to time in the universe as we know it, His present interventions are apparently limited to slowing down or temporarily arresting the process of decay of our physical world, not reversing it. I believe God controls all circumstances in life, such as hurricanes, earthquakes, airplane crashes, and the activity of the stock market. In fact, He "...accomplishes all things according to the counsel of his (own) will." (Ephesians 1:11) How He does this while still allowing man free will, how he takes into account all the thoughts and actions of several billion people including those yet unborn, is utterly amazing, of course.

Unfortunately, in spite of God's grace and mercy, man is increasingly suffering the disastrous consequences of his own bad moral choices. As has been often said, man does not violate the law of gravity by leaping off of a tall building; he merely illustrates it. It is perfectly obvious in the moral, social sphere that life on earth is not getting better and better, but worse and worse as history moves along! I think this disintegration and decay extends into our understanding of science, philosophy, history, art, and other areas of human endeavor as well. The late, eminent theologian Francis Schaeffer documented this theory eloquently in his many outstanding books on the history of civilization in the West.

# Temporary Energy Reversals in Some Miracles

We may, as individuals, have life easy for a time, but death catches us, each one, and the overall plight of mankind grows more perilous day by day. God may heal us when we become ill, but our restored good health may last only for a season. As I said, I do not know whether a local decrease in entropy such as we would expect from a "healing miracle of God" represents an input of new energy into the physical world, or simply a local energy transfer that does not change the sum total of the energy reserves of the Old Created Order. Lazarus was raised from the dead by Jesus, but he later died like everyone else in his generation. Did his temporary resurrection (actually a resuscitation) occur because new, outside energy was introduced into our physical universe, or did Jesus borrow the necessary energy from an existing energy reservoir in the old creation? However Jesus did it, it was meant to give us a picture in miniature of a greater and more permanent resurrection in which those raised would never die. Lazarus is not now with us, so the restoration of his life was temporary, the reversal of his death did not prevail. Yet we shall see Lazarus again in the great resurrection which is to come.

Every spring time, in temperate latitudes, new life bursts forth year after year after the winter snows have gone, leading many peoples around the world to think of "Mother Nature" as a living goddess who re-creates and renews the world seasonally and periodically. But green grass, buds, blossoms, and fruit are all brought into being by the energy of the sun acting on already-living cells with their existing genetic blueprints ever-ready - to reproduce, and grow, and mature given only favorable conditions. No force or energy source from outside the closed thermodynamic system of the present physical universe is necessary to explain how it all happens every springtime.

Of course, almost no one bothers to comment on the mystery of life itself! But we do all acknowledge that good physical health is often the result of living morally and righteously and showing respect for natural laws of diet, exercise, and rest. Conversely, when we violate moral law we can expect physical suffering - though not all suffering comes from that kind of causal source. It should be clear from God's revelation in nature, even apart from scripture, that there are laws in the spiritual realm that we do well to heed whether or not we know anything about physical laws and principles.

## Man Has Descended not Ascended

It is not hard to see that neither the Bible nor the Second Law of Thermodynamics adds any credibility to the popular myth of the evolution of life. The Second Law implies that the universe began with lower entropy than it now possesses, not the other way around. Not only is energy less and less available, but disorder and chaos are increasing with the passage of time. Man has not ascended but descended! Adam was more advanced than we, his sons. Civilization is largely an illusion, and except for God's intervention, many men live more like beasts nowadays than ever in the past! As for "ape men" and "cave men," we must see these aberrations as descendants of Adam who have wandered far from the truth and become degenerate and animal-like because God has given them up to their own self-destructive desires. They are not our ancestors, but sad and tragic sidebranches of mankind that give evidence of the severity of man's fall. (This assumes that the fossil record has been interpreted correctly in the first place! Many careful creation scientists have questioned whether ancient man really differed from modern man at all in such matters as brain size and fully human physical characteristics).

## An Old Creation and a New

The old creation led up to man as the peak, and the new creation leads down from Christ. The idea of an old created order of things (described in the Bible) is set in juxtaposition with the concept of a new creation now being quietly constructed as a home for God's new humanity: **"Therefore, if any one is in Christ, he is a *new creation*; the old has passed away, behold, the new has come."** (II Corinthians 5:17)

The "new creation" as described in the Bible is intimately tied to the resurrection out from among the dead of Jesus of Nazareth. Paul gives us a great wealth of information on this subject:

**"But in fact Christ has been raised from the dead, the first fruits of those who have fallen asleep. For as by a man (Adam) came death, by a man (Jesus) has come also the resurrection of the dead.**

49

For as (those who are) in **Adam all die, so also** (those who are) in **Christ shall all be made alive.** But each in his own order: **Christ the *first fruits*, then at his coming those who belong to Christ. Then comes the end, when he delivers the kingdom to God the Father after destroying every rule and every authority and power. For he must reign until he puts all his enemies under his feet. The last enemy to be destroyed is death.** 'For God has put all things in subjection under his feet (Psalm 8:6)' But when it says, 'All things are put in subjection under him,' it is plain that he is excepted who put all things under him. **When all things are subjected to him, then the Son himself will also be subjected to him who put all things under him, that God** (the Father) **may be everything to everyone."** (I Corinthians 15:20-18)

This passage gives the additional insight that the "new creation" is not totally separate and distinct from the old since Jesus is reigning now over the old creation and will renew it fully when He has destroyed out of it all his enemies. Furthermore, life in the new creation is born out of seeds sown into the soil of the old creation, figuratively speaking. The reference to "first fruits" clearly relates to the sheaf of grain offered by the Jews at the Feast of Unleavened Bread following the Passover, the very day on which Jesus was raised from the dead to become the Pioneer and Firstborn of the new creation.

## God Maintains and Controls the Old Creation

Another important claim of scripture about the old creation is that God is the present *Sustainer* of the universe. That is, He is not uninvolved, remote, detached and impersonal, leaving things to run by themselves by any means:

'In many separate revelations - each of which set forth a portion of the Truth - and in different ways God spoke of old to [our] forefathers in *and* by the prophets, [But] in the last of these days He has spoken to us in [the person of a] Son, Whom He appointed Heir *and* lawful Owner of all things, also by *and* through Whom He created the worlds *and* the reaches of space *and* the ages of time - [that is] [He made, produced, built, operated, and arranged them in order]. He is the sole expression of the glory of God - [the Light-being, the out-raying or radiance of the divine], - and He is the perfect imprint *and* very image of [God's] nature, upholding

50

*and* maintaining *and* guiding *and* propelling the universe by His mighty word of power..." (Hebrews 1:2-3) (Amplified Bible).

Again, the fact that Jesus is presently sustaining the universe from the realm of the spiritual raises the question whether there are inputs of spiritual energy into our physical world which ultimately show up as energy added "from the outside" of our physical world considered as a closed system. I do not know the answer to this question. It may be God sustains, directs, and propels the universe by operating only within the framework of the existing and known laws of physics we observe today. What I think I am safe in claiming for science is that no known violations of the Second Law have been detected (and verified) thus far by man. We have yet to learn whether miracles involve entropy changes introduced from outside the known (physical) universe, or whether UFO's are explainable entirely by presently understood rules of physics, and so on. As I mentioned in Chapter Two, it is hard for science to study rare or one-of-a-kind events, and such things tend to be relegated by default to the less reputable field of pseudoscience.

## Angels are Involved in God's Government

A Hebrew Old Testament concept, confirmed in the last book of the New Testament, the Book of the Revelation, is that *angels* have control over the forces of nature as instruments of God. The fact that angelic government and regulation of the sun, the winds, the rivers, and other forces of nature is uniform and unvarying means only that God and His servants are in perfect control, not that impersonal forces and fixed laws are *necessarily* behind the happenings of our physical world. Having set in motion the old universe, God **"rested on the seventh day,"** (Genesis 2:2) but that does not mean He withdrew from government of the old created order, as the opening verses of the letter of Hebrews show.

Since God gave Adam "dominion" over the original creation, one might well ask if God intended all of the "power flow" into nature (that is from the spiritual world into the physical) to come "through" man. One has only to look a little in the Bible to see that God governs and empowers nature (a) directly, (b) through natural forces, (c) through angels, (d) through men, and (4) supremely through His Son, who is the Lord of Creation.

The Book of Jonah contains several good illustrations of God's ability to directly command (and get an immediate response) from the winds so as to produce a great storm when needed, from a fish, a plant, and a worm. Man's response to God is sluggish and imperfect in comparison to nature's responses! While He walked the earth, Jesus had the power to directly command the forces of nature as well as demons and angels as presented frequently in the Four Gospels. Man's central role in God's purposes in creating the universe does come forth in the New Testament as well as the Old, supporting what modern secular cosmologists have come to consider "The Anthropic Cosmological Principle." The Biblical concept, however, is not that God is anthropomorphic but that man is theomorphic. The Apostle Peter says that men have been made, through redemption, **"...partakers of the divine nature."** (II Peter 1:4)

## Is Man the Center of the Universe?

It is a curious fact that man lies approximately midway between the size of the smallest atomic particle and the largest dimensions of the universe itself. It seems we have been placed squarely in the center! Man is a spirit who inhabits a body. The spirit gives man life and consciousness, and the physical body allows that spirit to have access to the physical world on the outside. As the spirit interacts with the body, the soul results. (The soul is the seat of the personality, emotions, the mind, the will, the memory, and so forth.)

Prior to coming into a personal relationship with Jesus Christ and experiencing spiritual regeneration, man is spoken of as "spiritually dead." "Spiritually dead" does not mean that some men have no spirits, for that state of affairs is synonymous with the death of the body. To be spiritually dead means to have a broken link, a missing connection with one's Creator - He alone has Life. Man in his "natural" state is spiritually dead and is motivated by "soulish" life and energy. Thus Paul writes concerning spiritual regeneration:

**"And you he made alive, when you were dead through the trespasses and sins in which you once walked, following the course of this world, following the prince of the power of the air, the spirit that is now at work in the sons of disobedience. Among these we all once lived in the passions of our flesh, following the**

desires of body and mind, and so we were by nature children of wrath, like the rest of mankind. But God, who is rich in mercy, out of the great love with which he loved us, even when we were dead through our trespasses, made us alive together with Christ (by grace you have been saved), and raised us up with him, and made us sit with him in the heavenly places in Christ Jesus, that in the coming ages he might show the immeasurable riches of his grace in kindness towards us in Christ Jesus. For by grace you have been saved through faith; and this is not your own doing, it is the gift of God-not because of works, lest any man should boast. For we are his workmanship, created in Christ Jesus for good works, which God has prepared beforehand, that we should walk in them." (Ephesians 2:1-10)

Whatever power and right man originally had to properly exercise control over creation, was largely lost at the fall. Only a partial restoration occurred when God called Adam and Eve back to Himself and made new provisions for their lives and for the creation. In giving their lives to the Evil One, our first parents abdicated, as it were, the title deed of the earth. At the fall, the alien, destructive power of Satan began to take control of man, devastating both people and nature. One of the aspects of Christ's work on the Cross was to ransom mankind and buy back from Satan any claims he may have had to dominion, territory or power in the universe. God's sovereign control over all creation - all the angels and all men - has, of course, never been diminished in the slightest or threatened by the presence of evil in the universe or by "plots against the throne." (I personally believe in the doctrine of "unlimited atonement" which states that Christ died for the sins of all mankind, not just for the sins of those who accept him as Lord of their lives. However, I am not a universalist. They believe that all men will ultimately be saved and none lost).

That our Creator originally intended man to have the central place in creation is clearly states by the writer of the letter to the Hebrews:

"For it was not to angels that God subjected the world to come, of which we are speaking. It has been testified somewhere, 'What is man that thou art mindful of him, or the son of man, that thou carest for him? Thou didst make him (man) for a little while lower than the angels, thou hast crowned him with glory and honor, *putting everything in subjection under his feet*.' Now in putting everything in subjection to him, he left nothing outside his control. As it is, we do not yet see everything in subjection to him.

53

"But we see Jesus, who for a little while was made lower than the angels, crowned with glory and honor because of his suffering of death, so that by the grace of God he might taste death for every one. For it was fitting that he, for whom and by whom all things exist, in bringing many sons to glory, should make the pioneer of their salvation perfect through suffering. Since therefore the children share in flesh and blood, he himself likewise partook of the same nature, that through death he might destroy him who has the power of death, that is the devil, and deliver all those who through fear of death were subject to lifelong bondage." (Hebrews 2:5-10; 14-15)

## Man, the Dwelling-Place of God

The New Testament takes care to refer to the body of the believer as the temple of God and to the church as a whole as a building of living stones built upon Christ the Cornerstone. Thus, the tabernacle of Moses in the wilderness and the two Jewish temples that followed in Jerusalem were pictures of man. The tabernacle had an outer court, a holy place, and a most holy place, each with specific items of furniture depicting man's approach to God through suitable sacrifice, confession of sin, purification from sin, enlightenment into correct knowledge, and prayer. The fact that the Bible describes man as the dwelling place of God is emphasized in a number of passages in the Bible, but most people imagine that - if there is a God - He lives outside the known universe and rules as a detached observer intervening only rarely in the affairs of the world. The Biblical picture is that God indwelt man in the Garden of Eden and empowered man. The fall was man's abdication of his exalted position as God's steward over all He had made. Jesus, however, the Son of man, is the faithful steward and rightful heir of all things. Those who belong to Christ Jesus are indwelt once again by Him, by the Holy Spirit, and this places man back at the center of things as God's temple, God's dwelling place. These two great truths are emphasized in the New Testament: "Christ is in us" (Colossians) and "We are in Christ" (Ephesians).

In his Second Epistle Peter tells us more about God's present "energetic" participation in the old created order:

"But the day of the Lord will come like a thief, and then the heavens will pass away with a loud noise, and the elements will be dissolved (*unloosed*) with fire, and the earth and the works that are upon it will be burned up." (II Peter 3:10)

This passage suggests that the active power of God is responsible for such things as the nuclear binding energy, the "strong force," in every atom - without which the atomic nucleus would fly apart. The popular gospel chorus which assures us that "He's got the whole, wide world in His hands" is perfectly correct theologically. Scripture assures us (if I may reiterate) that volcanoes do not erupt, tornadoes do not wreak their havoc, and sparrows do not fall to the ground outside of God's control - regardless of the intermediate mechanisms by which He regulates these activities and happenings of the physical world. Though it may appear so to us, there are really no accidents in a universe where a living God pays perfect attention to every detail and loses sight of nothing:

"O LORD, how manifold are thy works! In wisdom hast thou made them all; the earth is full of thy creatures. Yonder is the sea, great and wide, which teems with things innumerable, living things both small and great. There go the ships, and Leviathan which thou didst form to sport in it. These all look to thee, to give them their food in due season. When thou givest to them, they gather it up; when thou openest thy hand, they are filled with good things. When thou hidest thy face, they are dismayed; when thou takest away their breath, they die and return to their dust. When thou sendest forth thy Spirit, they are created; and thou renewest the face of the ground." (Psalm 104:24-30).

A clear distinction is made in Holy Writ between the Creator and the created thing. By contrast, pantheistic religious views such as Hinduism, hold that the sum total of all things is God. The God of the Bible stands apart from His creation. But He is also actively involved in the sustenance of his creatures and the establishment of an invisible kingdom which is quietly being prepared for public unveiling at the end of our age.

## A Coming New Creation

Clues to the nature of the "New Creation" are given in the remaining portions of the fifteenth chapter of First Corinthians:

"But some one will ask, 'How are the dead raised? With what kind of body do they come?' You foolish man! What you sow does not come to life unless it dies. And what you sow is not the body which is to be, but a bare kernel, perhaps of wheat or of some other grain. But God gives it a body as he has chosen, and to each kind of seed its own body. For not all flesh is alike, but there is one kind for men, another for animals, another for birds, and another for fish. There are celestial bodies and there are terrestrial bodies; but the glory of the celestial is one, and the glory of the terrestrial is another. There is one glory of the sun, and another glory of the moon, and another glory of the stars; for star differs from star in glory.

"So it is with the resurrection of the dead. What is sown is perishable, what is raised is imperishable. It is sown in dishonor, it is raised in glory. It is sown in weakness, it is raised in power. It is sown a physical body, it is raised a spiritual body. If there is a physical body, there is also a spiritual body. Thus it is written, 'The first man Adam become a living being'; the last Adam became a life-giving spirit.

"But it is not the spiritual which is first but the physical, and then the spiritual. The first man was from the earth, a man of dust; the second man is from heaven. As was the man of dust, so are those who are of dust; and as is the man of heaven, so are those who are of heaven. Just as we have borne the image of the man of dust, we shall also bear the image of the man of heaven. I tell you this, brethren: flesh and blood cannot inherit the kingdom of God, nor does the perishable inherit the imperishable. Lo! I tell you a mystery. We shall not all sleep, but we shall all be changed, in a moment, in the twinkling of an eye, at the last trumpet. For the trumpet shall sound, and the dead will be raised imperishable, and we shall be changed. For this perishable nature must put on the imperishable, and this mortal nature must put on immortality. When the perishable puts on the imperishable, and the mortal puts on immortality, then shall come to pass the saying that is written: 'Death is swallowed up in victory.' 'O death, where is thy victory? O death, where is thy sting?'"
(I Corinthians 15:35-55)

In this passage we see evidence that the universe is more open-ended than we might otherwise suppose. There are several "levels" to creation, and not all living beings are "creatures made of dust" as we are. The angels are made entirely of spirit for instance, as far as I know. To move into the new creation, one must die. The body is planted like a seed, but the new life which springs forth is altogether different from the seed, though derived from the seed. God allows us to die, as he allows the old creation to "run down," but he then brings forth a new kind of body to clothe the redeemed of earth - those previously made new in their souls and spirits. Accompanying the resurrection is a renewal of creation and a releasing of nature from its bondage to decay. Second Corinthians 5:1-5 indicates the new resurrection bodies are already in existence waiting to be stepped into on the day of resurrection, or at the hour of one's death.

Jesus told his disciples He was going to prepare a place for them to live, which is now already in existence. Hebrews, Chapter 12, describes believers coming to New Jerusalem, which is the heavenly city now ready, and Revelation 21 describes that city in detail, like an orbiting space station of awesome beauty made of rare and costly materials prepared as a dwelling place for the new mankind. These matters are the subject of Chapter 12.

## God Uses Natural Forces to Control Many Events

That God should use "natural" processes to feed and provide for us in this life as we live out our lives in the old creation makes Him no less personal. As discussed, if God intervenes now in the old creation by perturbing the "laws of nature" or introducing new energy from outside, He apparently does so in modest ways. As I have emphasized, God does however fully control human events, **"...accomplishes all things according to the counsel of his will."** (Ephesians 1:11) In the New Creation the processes of death and decay now operative in our world will be undone and turned around. Disorder and disharmony will be reversed and a whole new order of things will permeate the old. Thus, (it seems to me) that the so-called immutable laws of physics are really only applicable to a certain time frame of God's dealings with men and also to a particular limited realm of the Old Creation. They do not necessarily describe the entire time period since creation, but perhaps are special cases only.[3]

The Bible claims that our Creator will intervene on earth one day soon, destroying the old creation and its works and completing the new creation which has already begun with the transformation of the lives of believers.

## Jerusalem, The Center of the New Creation

"For behold, I create new heavens and a new earth; and the former things shall not be remembered or come into mind. But be glad and rejoice for ever in that which I create; for behold, I create Jerusalem a rejoicing, and her people a joy. I will rejoice in Jerusalem, and be glad in my people; no more shall be heard in it the sound of weeping and the cry of distress. No more shall there be in it an infant that lives but a few days, or an old man who does not fill out his days, for the child shall die a hundred years old, and the sinner a hundred years old shall be accursed. They shall build houses and inhabit them; they shall plant their vineyards and eat their fruit. They shall not build and another inhabit; they shall not plant and another eat; for like the days of a tree shall the days of my people be, and my chosen shall long enjoy the work of their hands.

"They shall not labor in vain, or bear children for calamity; for they shall be the offspring of the blessed of the LORD, and their children with them. Before they call I will answer, while they are yet speaking I will hear. The wolf and the lamb shall feed together, the lion shall eat straw like the ox; and dust shall be the serpent's food. They shall not hurt or destroy in all my holy mountain, says the LORD." (Isaiah 65:17-25)

## Notes to Chapter Three

1. I highly recommend Charles B. Thaxton, Walter L. Bradley and Roger L. Olsen, *The Mystery of Life's Origin* (Philosophical Library; New York, 1984), R. L. Wysong's *The Creation-Evolution Controversy* (Inquiry Press; 4925 Jefferson Ave.; Midland, Michigan 48640; 1976), and Michael Pitman, *Adam and Evolution* (Rider and Company; London, 1984). These books concern the genetic-coding mechanisms in living things and the impossibility of self-assembly of such complex structures by a combination of time plus chance. Because statistics are so overwhelmingly against the spontaneous self-assembly of complex organic molecules - let alone living cells - the idea that life originated in outer space and came here via meteor or comet (the "panspermia" theory) has gained popularity in recent years.

Of course, the theory begs the question as to how life began in outer space where the same laws of physics prevail. See Fred Hoyle, *The Intelligent Universe* (Holt, Rinehart, and Winston; New York, 1980), for this view.

2. The Bible indicates there are various levels of life possessed by plants, by animals, by men and by angels. (I Corinthians 15)

3. C. S. Lewis in his *Mere Christianity* discusses the multi-dimensional nature of time and space in clear terms that are most helpful. The laws of physics may be the same on the farthest star as they are on earth but the visible universe, vast though it seems is only a small corner of creation, as the Bible makes clear.

# Chapter Four

## The Steps of Creation

### Inadequacy of the Big Bang Hypothesis

For the past five decades or so, the prevailing scientific explanation of the origin of the universe has been the "Big Bang" hypothesis, which became popular after it was discovered that light from distant stars and galaxies was shifted down the spectrum towards the red. The more distant the object, the greater the shift, apparently. The "red shift" was assumed to be really only the familiar Doppler shift of light or sound emitted from a moving object. A common example of the Doppler shift is the changing pitch of a train whistle as it approaches, passes, and recedes from an observer on the station platform. Other explanations for the red-shift have been proposed, but until now, none have enjoyed much popularity.

According to the Big Bang theory, the universe began, essentially, with a burst of intense light. The theory is quite an ingenious one and has been worked out in greater detail[1] including many computer calculations of the early history of the universe before matter is supposed to have condensed out of pure radiation, when temperatures were too high and densities too great for matter as we know it now to exist. Although there is much to commend the Big Bang model, serious flaws have begun to appear in recent years because of new evidence from the farthest reaches of space and new questions raised in nuclear physics. Also, as we shall see, this theory does not fit the Biblical account because it is too limited in scope.

As we have already noted, if the Big Bang process had been the way the universe came into being, then apart from the ongoing involvement of an outside Intelligence to bring order out of chaos, the entropy of the universe was at its lowest value at the moment of creation. The initial burst of light of the "Big Bang" had to contain the full wealth of programing information that would later lead to all the complexities we find around us now in nature and in ourselves. Either entropy was initially low and the "assembly program" was built in from the start, or "God" molded and fashioned the universe as it unfolded. In the latter case God must have intervened continuously, or at least intermittently, altering the now-fixed "laws of physics" after the big-bang occurred.

## Many Secrets of Creation *Are* in the Bible

To understand creation more fully, we are given more than we might first suppose in Chapter One of Genesis and elsewhere in bits and pieces in the Bible. For instance Genesis, 1:2 *does* imply that God first brought the universe into being, out of nothing and that then he sculpted, formed, and fashioned it like an artisan at work on raw material: **"The earth was without form and void, and darkness was upon the face of the deep; and the Spirit of God was moving over the face of the waters."** (Genesis 1:2)

The narrative has already centered on the earth (as opposed to the heavens) between the first and second verse of Genesis, and although we could wish that God had told us much more, the entire creation account in Genesis occupies only 0.3% of Holy Writ! Evidently God has more important things He wishes us to know than *exactly* how he brought things into being!

I do want to emphasize again that the Bible is an unusual book in that material on any one given subject is not all found in the same book or chapter or section of the Bible. The Bible teaches us about all of life, **"...precept upon precept, precept upon precept, line upon line, line upon line, here a little, there a little."** (Isaiah 28:13) To be well-balanced Christians, we must read the entire Bible and not just selected sections. It is also unwise to become a specialist in creationism or eschatology, or "victorious Christian living" to the neglect of other subjects. The need for a balanced diet and a balanced life applies to more than food and exercise! Paul the Apostle having spent many months teaching at Ephesus, told the elders of the church there, as he left them for the last time, that he **"...did not shrink from declaring to you the whole counsel of God."** (Acts 20:27)

It should not be necessary to dwell on the fact that knowledge of the things of God is hindered and even lost by hypocrisy or lack of righteous conduct. Among professing Christians in the church, many are not regenerated, and, of course, their presence, influence, and teaching will be in error. Yet they may not be recognized as representing ideas foreign to the thoughts and heart of God, his ways, and his purposes among men. Whole churches have completely departed from the faith in this century, so that the name "Christian" or the beliefs they once cherished now mean nothing. Compounding the situation is the widespread biblical illiteracy which is commonplace today.

## Dimensions of Time in Creation

We are accustomed to thinking of a linear cause-and-effect relationship between events. However, numerous Bible commentators have noted that in the six days of creation the sequence of events in the creative activity of God, does not occur in a strict linear sequence in time. Often in the Bible we see that some single creative event in the spiritual realm simultaneously affects events in the physical world that are not intimately connected with one another. In Genesis, the first and fourth, second and fifth, and third and sixth days go together:

Day One:
**"And God said, 'Let there be light'; and there was light. And God saw that the light was good; and God separated the light from the darkness. God called the light Day, and the darkness he called Night. And there was evening and morning, one day."** (Genesis 1:3-5)

Day Four:
**"And God said, 'Let there be lights in the firmament of the heavens to separate the day from the night; and let them be for** (marking out, measuring) **signs and for seasons and for days and for years, and let them be lights in the firmament of the heavens to give light upon the earth.' And it was so. And God made the two great lights, the greater light to rule the day and the lesser light to rule the night; he made the stars also. And God set them in the firmament of the heavens to give light upon the earth, to rule over the day and over the night, and to separate the light from the darkness. And God saw that it was good. And there was evening and there was morning, a fourth day."** (Genesis 1:14-19)

✦

Day Two:
"And God said, 'Let there be a firmament in the midst of the waters, and let it separate the waters from the waters.' And God made the firmament and separated the waters which were under the firmament from the waters which were above the firmament. And it was so. And God called the firmament Heaven. And there was evening and there was morning, a second day." (Genesis 1:6-8)

Day Five:
"And God said, 'Let the waters bring forth swarms of living creatures, and let birds fly above the earth across the firmament of the heavens.' So God created the great sea monsters and every living creature that moves, with which the waters swarm, according to their kinds, and every winged bird according to its kind. And God saw that it was good. And God blessed them, saying, 'Be fruitful and multiply and fill the waters in the seas, and let birds multiply on the earth.' And there was evening and there was morning, a fifth day." (Genesis 1:20-23)

Day Three:
"And God said, 'Let the waters under the heavens be gathered together into one place, and let dry land appear.' And it was so. God called the dry land Earth (*eretz*), and the waters that were gathered together he called Seas. And God saw that it was good. And God said, 'Let the earth put forth vegetation, plants yielding seed, and fruit trees bearing fruit in which is their seed, each according to its kind, upon the earth.' The earth brought forth vegetation, plants yielding seed according to their own kinds, and trees bearing fruit in which is their seed, each according to its kind. And God saw that it was good. And there was evening and there was morning a third day." (Genesis 1:9-13)

Day Six:
"And God said, "Let the earth bring forth living creatures according to their kinds: cattle and creeping things and beasts of the earth according to their kinds.' And it was so. And God made the beasts of the earth according to their kinds and the cattle according to their kinds, and everything that creeps upon the earth according to its kind. And God saw that it was good. Then God (*Elohim*) said, 'Let us make man in our image, after our

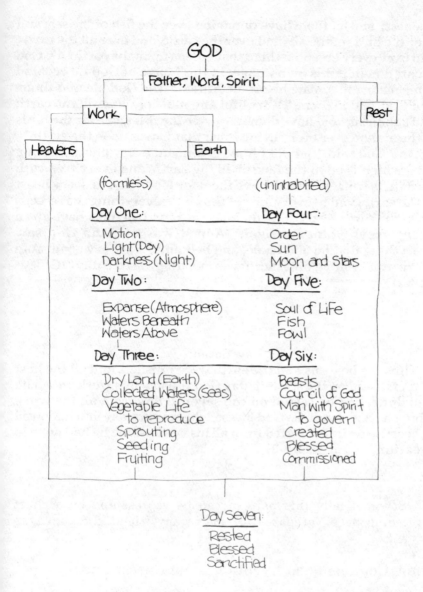

GOD

Father, Word, Spirit

Work

Rest

Heavens

Earth

(formless)

(uninhabited)

Day One:

Day Four:

Motion
Light (Day)
Darkness (Night)

Order
Sun
Moon and Stars

Day Two:

Day Five:

Expanse (Atmosphere)
Waters Beneath
Waters Above

Soul of Life
Fish
Fowl

Day Three:

Day Six:

Dry Land (Earth)
Collected Waters (Seas)
Vegetable Life
    to reproduce
Sprouting
Seeding
Fruiting

Beasts
Council of God
Man with Spirit
    to govern
Created
Blessed
Commissioned

Day Seven:

Rested
Blessed
Sanctified

likeness; and let them have dominion over the fish of the sea, and over the birds of the air, and over the cattle, and over all the earth, and over every creeping thing that creeps upon the earth.' So God created man in his own image, in the image of God he created him; male and female he created them. And God blessed them, and God said to them, 'Be fruitful and multiply, and fill the earth and subdue it; and have dominion over the fish and over the birds of the air and over every living thing that moves upon the earth.'

"And God said, 'Behold I have given you every plant yielding seed which is upon the face of all the earth, and every tree with seed in its fruit; you shall have them for food. And to every beast of the earth, and to every bird of the air, to everything that creeps upon the earth, everything that has the breath of life, I have given every green plant for food.' And it was so. And God saw everything that he had made, and behold, it was very good. And there was evening and there was morning, a sixth day." (Genesis 1:24-31)

Day Seven:
"Thus the heavens and the earth were finished, and all the host of them. And on the seventh day God finished his work which he had done, and he rested on the seventh day from all the work which he had done. So God blessed the seventh day and hallowed it, because on it God rested from all his work which he had done in creation." (Genesis 2:1-3)

Schematically this pattern can be represented by a chart (preceding page), prepared for me by my friend and colleague Bryce Self.

## Some Comments on Events During the Seven Days

A few brief comments on specifics of the Seven Days of Creation will help set the stage for some of the important points I have set out later on in this book. On Day Two we note that the sun, moon, and stars and their motions in the heavens with respect to the earth are given to man for measuring time and recording events. Thus,

Scripture establishes "gravity clocks" or "dynamical time" as the Biblical way of keeping track of time. That is, the rotation of the earth on its axis measures the length of the day; the revolution of the moon around the earth establishes the month; the revolution of the earth about the sun measures the year; the constellations mark the seasons; and the planets, comets and special stars signal special events such as the birth of the Messiah, Jesus. The rate of movement of planets, pendulum clocks, water clocks, and sand hour-glasses depends on the *gravitational constant*, G, known to the physicist.

A more thorough discussion of time is given in Chapter Five. We will look into gravity and gravity clocks more carefully in Chapters Six and Seven. Geological time is commonly measured by the "clock" of the radioactive decay rates of certain atomic nuclei. As we shall see, radioactive clocks have apparently been running at a different rate compared to gravity clocks since very early times. History is marked out in the Bible by epochs, dynasties, and so-called "dispensations," (when God changes His priorities and emphases in the government of the nations).

There is also a different kind of time in the spiritual realm of the heavens which apparently does correspond (but not in a one-to-one manner) with time intervals as measured on earth. In "eternity" there may well be such a thing as "timelessness"; however, all meaningful human events require a succession of events - that is, a progression of time. In the Bible we are given a number of glimpses into heaven where we see sequences of events taking place activities of God and the angels, for instance, that are related to, but not one-on-one with, events transpiring on earth.

The New Testament says **"God is Light"**. And the Apostle John (in the prologue to his Gospel and in his First Epistle) contrasts light and darkness as moral: **"...God is light and in him there is no darkness at all."** (I John 1:5) Nighttime is not evil, and daytime good, of course, however the Genesis record is describing for us created things in the heavens that correspond to things on earth, both at the same time. Just as there is physical light,[2] there is also spiritual light underlying and energizing physical light. Spiritual light is the source, physical light is the derivitive. The entire creation as originally made by God was pronounced by Him as "good" - all of it, including night as well as day. The stars are symbols for the angels. The sun is a picture of Christ "who rules the Day," (that is, when He is physically present on the earth). While the moon, is the lesser light shining only by reflected light from the sun, "rules the night," (that is, during the present age when Christ is not physically, visibly present with us). For example, certain false teachers are characterized by Jude as, "**...wandering stars for whom the nether gloom of darkness has been reserved forever.**" (Jude 13)   Paul says "**...even Satan disguises himself an angel of lght.**" (II Corinthians 11:14)   In contrast, Jesus said, "**...I am the light of the world;  he who follows me will not walk in darkness, but will have the light of life.**" (John 8:12)

The English word "firmament" (Day Two) does not adequately reflect the Hebrew *raqia* - which comes from a verb meaning "to hammer out" (as a goldsmith hammers gold leaf), or "to stretch out." (See also Psalm 104:2, Isaiah 40:22, 42.5, and 44:24). The idea here is the initial stretching out of the fabric of space itself, which (as we shall see later) causes "empty" space to have certain physical properties.   "Empty" space has both electrical and gravitational properties known as "metrics" which determine in effect the "scale" of space-time.

Some capable physicists in our time, whom I respect, have renewed the concept of an "ether" as an invisible medium which constitutes the fabric of space.[3] The majority of physicists today have abandoned the idea that an ether exists.  In either case, whether one assumes there is a carrier medium or simply the vacuum of empty space, space itself is known to have certain mechanical and electrical properties which can be variable. C. S. Lewis conceives of the planets as occupying holes in space in his science fiction trilogy. That is, he inferred that space has inherent fullness in comparison to matter.  I suspect this is a realistic way of looking both at the world of the spirit and the "vacuum" of space.

In regard to the formation of dry land in the midst of the waters (Day Three), Peter in his Second Epistle describes the earth as originally being "**...formed out of water and by means of water.**" (II Peter 3:5) Thirty years ago astronomers held that the earth began as a molten blob that cooled, with gases and water boiling out of the hot interior later on.  However, the evidence now in hand has led modern astronomers in recent years to conclude that the earth and planets formed by a cold accretion process (that is, built up by a gathering together of dust, debris, and small objects) with abundant trapping of water and gases in the mantle. To be sure, the interior of the earth is hot now, but, as we shall see, most of this heating has come about by radioactive decay of unstable atomic elements in the interior of the earth.

Consistent with the modern theory of continental drift, the Bible indicates that there was originally but one land mass on the earth, surrounded by water. Later in this book I will discuss the possibility that the splitting apart of the one continent into many took place rapidly in time as a result of a meteor or comet impacting the earth about 2345 B.C.

In the creation account, seven times God observes that what He had made was "good." The last time, He declares that it was "very good." I suggest that the present creation has been badly marred and flawed since the fall of man (and the angels). These fall(s) have affected not only man but also nature and the laws of physics as we know them today, as I will attempt to show as we progress.

The "hosts" mentioned at the end of the narrative on creation include the numerous angels as well as the various populating species of the earth. The angels were created at *some* time prior to the seventh day. The number of angels is very great. Revelation comments on "myriads of myriads" and since a myriad is 10,000, the total number of angels evidently exceeds 100 million!

## God Creates from Outside of Time

The creative activity of God during the six days of creation evidently took place from outside of physical, dynamical time. However, all of the creative activity of God took place during *an interval of time*, (six days), whether we measure time by a spiritual clock (a clock that would tell us the sequence of events in the heavens) or by a physical clock. The creation of the universe and all that fills it had a definite starting point, and after a sequence of creative acts over a span of time, God had completed the creation of the universe. Thereafter, He did not do any additional "creating" but allowed the universe to run according to pre-determined and built-in laws. (This is not to deny God's ongoing involvement as the Sustainer and Lord of Creation).

We are only able to read books a line at a time, and we are accustomed to a strictly linear flow of events. It is difficult for us to accommodate the notion of God bringing into existence today things that we shall not be able to see, measure, or experience until tomorrow. It is hard for us to grasp the notion that everything is "now" to God: the past, the present, and the future as well. Creation from outside of time is much more complicated than one first supposes,[4] and it is difficult to decide how time is being measured to an imaginary human observer who might have been standing on earth during "creation week." Some argue for literal 24 hour days measured by God's Grandfather's Clock at the center of the universe; others insist the "days" are long periods of time.

## A Different Quality of Time Before the Fall

I personally think that *time as we know it now* began with the fall of man and that time had different "qualities" before the fall. If the day is defined as one rotation of the earth about its axis, then it is possible the length of that day has changed somewhat from its initial value, but surely not by orders of magnitude (i. e., by factors of ten). On the other hand, the subjective time scale of Adam's life may have contained the quantity and quality of experience we would scarcely accumulate in many years.

The fact that the creative activity of God took place in a time sequence from Day One through Day Seven indicates that time was "flowing" in its usual sense from past through present towards the future. Yet as the above regrouping of the Days (1-4, 2-5, 3-6) shows, there is an "eternal" dimension also present all through creation week. Neither Satan nor Adam had fallen, there was as yet no sin. "Perfect physics" prevailed, and the spiritual dimension of the universe was "in tune" with the physical in a way we cannot now exactly understand.

## The Time of Origin of the Angels?

The brevity of the Genesis account raises in my mind a number of hard questions. For example, when were the angels created? Some Bible scholars say the angels were created on Day Five because the stars were made on the Fifth Day and angels in Scripture are closely associated with the stars. Also, the created universe begins with the simple and moves towards the complex. The creation of man on the Sixth Day, after all the rest of the ecological environment was ready and functioning, suggests that the angels might have been created next-to-last, as great beings whom God uses as messengers and agents in his government of the universe. Since the universe was made for man, it would be logical to introduce the angels just prior to man's arrival, as the final setting of the stage for the man of dust who was to be formed into the image and likeness of his Creator. The creation of the angels on the same day the stars were created seems to some scholars to be supported by Psalm 148:

Praise the LORD!
Praise the LORD from the heavens,
praise him in the heights!
Praise him, all his angels,
praise him, all his host!
Praise him, sun and moon,
praise him, all you shining stars!
Praise him, you highest heavens,
and you waters above the heavens!
Let them praise the name of the LORD!
For he commanded and they were created.
And he established them for ever and ever;
he fixed their bounds which cannot be passed.
Praise the LORD from the earth,
you sea monsters and all deeps,
fire and hail, snow and frost,
stormy wind fulfilling his command!
Mountains and all hills,
fruit trees and all cedars!
Beasts and all cattle,
creeping things and flying birds!
Kings of the earth and all peoples,
princes and all rulers of the earth!
Young men and maidens together,
old men and children!
Let them praise the name of the LORD,
for his name alone is exalted;
his glory is above earth and heaven.
He has raised up a horn for his people,
praise for all his saints,
for the people of Israel who are near to him.
Praise the LORD!

On the other hand, the angels may have been created before the First Day. This can be supported by a verse in the book of Job which suggests the angels observed the creation of the rest of the universe as bystanders. God questions Job:

"**Where were you when I laid the foundation of the earth? Tell me, if you have understanding. Who determined its measurements-surely you know! Or who stretched the line upon it? On what were its bases sunk, or who laid its cornerstone, when the morning stars sang together, and all the sons of God shouted for joy?**" (Job 38:4-7)

Since "**the sons of God**" in the opening verses of Job are most certainly the angels, this passage is often taken as evidence that the angels were created early on, included in the general statement, "**In the beginning God created the heavens and the earth.**" (Genesis 1:1) At any rate, by the seventh day not only the heavens and the earth had been created, but all the "hosts" that fill them, and that includes the angels, (Genesis 2:1).

It would be helpful (I think) to know when the angels were created, but God evidently does not think it that important to tell us for certain. Among both Jewish and Christians scholars we find divided schools of thought on this question. However, we can be certain that the fall of Lucifer (Satan), one of the mightiest angels before God, occurred before the fall of man since he tempted and seduced Adam and his wife Eve in the garden.

## When Did Eve Appear?

Another hard question is how long did Adam remain alone in the garden before Eve was created from his side? To ask about time when the scene is taking place in a realm of time different from that which we now experience may be meaningless. Some say Eve was created a little later on during the "first" Sixth Day. If, as many suppose, the days of creation begin with evening, then perhaps Eve could have shown up sometime early in the morning on the first Friday.

However, God gave Adam the assignment of exploring the garden and naming all the animals before Eve was presented to him. Since Adam had to study, analyze, and understand thousands of species (many more than now exist!), surely this assignment required more than a couple of hours' efforts even for a sinless man!

# How Much "Time" Before the Fall?

How long did Adam and Eve remain in the garden enjoying its fruits and delights and rich fellowship with God before the fall? Some commentators think the fall occurred on the Sixth Day also, which means that the man and his wife must have been expelled before the first Sabbath, that is, prior to sundown on the first Friday!

Actually, at the end of the Sixth Day God pronounced everything "very good" so that *the fall could not have happened on the first of the Sixth Days* but must have occurred at some time after the first week. Therefore, weeks or even years of "time" could have elapsed before the fall. This time interval would have been subjectively different from time as we now experience it, though I suppose one could keep track of it by counting the earth's rotations on its axis. Adam and Eve were to be fruitful and to multiply. During this time they would surely "grow" but not grow older, that is they would not become more senile, which is to experience decreasing vitality and strength leading to death. Yet they bore no children before the fall, and one would suppose the delights of their sexuality were something they began to explore soon after they were presented to one another. They had not had time to eat of the Tree of Life which was not forbidden them.

Surely before the fall, the garden in which they were placed must have been like Paradise and the quality of a single "day" there rich and eternal, full of content and meaning. A verse in Psalm 90 quoted by the Apostle Peter probably addresses the richness of time in a world where the spiritual and the physical aspects of creation are in close harmony one with another. Peter says, **"...with the Lord one day is as a thousand years, and a thousand years as one day."** (II Peter 3:8) It is death that causes boredom and fatigue and robs life of its meaning. In an eternal setting where no sin is present, "clock time" may have little meaning compared to how we understand it now.

I do not know the answers to these questions, and I distrust simple answers that are attempts to extrapolate backwards from present conditions of time and space as we know them. I feel certain that the creation had at least an appearance of age, moments after it was brought into existence. Perhaps even mature trees had growth rings and the animals were fully grown when they appeared on the scene.[5]

Creation scientists much more learned and experienced than I have suggested such ideas for many years.

## Disasters After the Creation of all Things

Two great, terrible disasters have come upon the universe since God rested on the Seventh Day from all the works that He had made. These disasters: the fall of Lucifer and the fall of man, have so radically disrupted mankind, and nature, and the physical world, we can hardly hope to recapture what it must have been like before evil entered in. We cannot, therefore, settle definitely a number of critical issues about what life was like before the fall. Time as it was experienced by man before his fall may well not be definable or measurable by the kind of clocks and standards we now use. This I take as my present working hypothesis.

"Rejoice in the Lord, O you righteous!
Praise befits the upright.
Praise the LORD with the lyre,
make melody to him with the harp of ten strings!
Sing to him a new song, play skillfully on the strings,
with loud shouts.
For the word of the LORD is upright;
and all his work is done in faithfulness.
He loves righteousness and justice;
the earth is full of the steadfast love of the LORD.
By the word of the LORD the heavens were made,
and all their host by the breath of his mouth.
He gathered the waters of the sea as in a bottle;
he puts the deeps in storehouses.
Let all the earth fear the LORD,
let all the inhabitants of the world stand in awe of him!
For he spoke, and it came to be;
he commanded, and it stood forth.
The LORD brings the counsel of the nations to nought;
he frustrates the plans of the peoples.
The counsel of the LORD stands for ever,
the thoughts of his heart to all generations.
Blessed is the nation whose God is the LORD,
the people whom he has chosen as his heritage!
The LORD looks down from heaven,

74

he sees all the sons of men;
from where he sits enthroned he looks forth
on all the inhabitants of the earth,
he who fashions the heart of them all,
and observes all their deeds.
A king is not saved by his great army;
a warrior is not delivered by his great strength.
The war horse is a vain hope for victory,
and by its great might it cannot save.
Behold, the eye of the LORD is on those who fear him,
on those who hope in his steadfast love,
that he may deliver their soul from death,
and keep them alive in famine.
Our soul waits for the LORD;
he is our help and shield.
Yea, our heart is glad in him,
because we trust in his holy name.
Let thy steadfast love, O LORD, be upon us,
even as we hope in thee."
(Psalm 33)

## The Complexity of Persons in The Godhead

Were it not for God's revelation of Himself, we would surely invent all sorts of substitutes. In fact, this has been done by mankind anyway! Not only is our modern world full of idolatry, but world religions and pseudo-Christian cults have distorted the Biblical view of God showing Him to be different from who He really is.

All three persons of the Godhead: Father, Son, and Holy Spirit are at one time or another called "God" or "Lord" in the New Testament, and all demonstrate the powers and attributes of God. The Bible speaks, however, of one God, not three, and the mystery of the interrelationships between the three Persons is ultimately beyond our comprehension - the creature cannot exhaustively analyze and know all there is to know about its Creator! However, the Three Persons of the Godhead counsel together and then act in creating the world and in redeeming mankind, and they evidently work together by an amazing process of mutual self-giving love.

75

Genesis assures us that man was created in the image and likeness of God. By "man" Scripture means "man and woman," Adam and Eve. Just as many scientists ignore revelation, assuming they have all the truth in hand, so also others of us easily forget the importance of a woman's point of view. The fact that Eve in her femininity carries one-half of the image of God is all too easily forgotten. Western philosophy tends to demean the East, yet both have something important to say about the nature of reality. Thus holding complementary points of view in tension together is often necessary if we are to understand reality.

## A Fourth Person in the Trinity?

In 1937 while lecturing at Yale University,[6] the late C. G. Jung raised some old issues related to the nature of God and man's relationship with God that are popular today in certain circles, Although Jung was thoroughly acquainted with Christianity and with Eastern religions and wrestled with the problem of human evil, there is no evidence he himself ever became a Christian. More properly he should be thought of as a modern gnostic. However, not all his insights into the human psyche are incorrect, by any means, and not all his questions about the nature of God were naive. Jung took note that medieval philosophers felt the notion of a fourfold symmetry seemed in some ways to typify man's relationship to God better than the notion of the Trinity. Had these ancients been correct, we would today worship a Quaternity rather than the Trinity! It was the desire of Jung as well as these medieval alchemists to find a Fourth Person in the godhead - SomeOne personifying the Feminine.

The notion that God is One and indivisible is central to Judaism although the plural masculine noun *Elohim* (which usually takes a singular verb) implies "three or more." The New Testament never uses the term "trinity" or "triune" when referring to God; each of these three persons is always called by the masculine pronoun "He." In spite of attempts to change the genders of male and female in some new Bible translations, there is little doubt that God's basic, plain revelation of Himself is that He is but One God eternally-existing in Three Persons. God is a Spirit, of course, and ordinarily we do not think of spirits as having gender. God's revelation of Himself as Yahweh, "I am who I am", could refer to either masculinity or femininity, or both.

The problem of God's "gender identity" arises early in the Bible, for in Genesis Elohim says, "...**Let us make man in our image, after our likeness; and let them have dominion over the fish of the sea, and over the birds of the earth, and over the cattle, and over every creeping thing that creeps upon the earth. So God created man in his own image, in the image of God he created him: male and female he created them.**" (Genesis 1:26-27)

At this point in the narrative, we understand that "man" as created included "woman." That is, the first man, Adam, was actually Adam/Eve. Only later was Eve taken out and presented to Adam as a separate, differentiated being. We immediately understand that God Himself is not a sexual being as we are, but that He is equally masculine and feminine - as much a Mother as a Father. Sexuality as we know it arises out of the polarization of Adam and Eve, brought about after the creation of man, and the subsequent strivings of the two to become one again.

One attempt to solve the problem that God must somehow contain the feminine was the early Roman Catholic doctrine known as the "Assumption of the Virgin Mary." This doctrine was made an official Church Dogma on November 1, 1950 by Pope Pius XII. It is a statement of belief that Mary, the Mother of Jesus, was taken up ("assumed") into heaven at the end of her life. There she became Queen of Heaven and "Co-Mediatrix" with Christ, hearing the prayers of the faithful and interceding with her Son on their behalf. Some in the Eastern Orthodox Churches added to this belief that Mary was not merely the mother of the humanity of our Lord Jesus, but was herself immaculately conceived (without sin) and is actually the "Mother of God" (**Theotokos**). This suggests that she has been part of the godhead from the beginning.

None of these notions can be supported from the Bible; in fact, they seem to have been borrowed from the pagan religion of the "Great Mother" whose roots are in the Babylonian Mystery Religion[7] founded by Nimrod after the flood. Certainly Mary deserves our respect and honor as a devout Jewish girl who gave herself to the purposes of the God of Israel that the Messiah might be brought forth into the family of Israel, in fulfillment of ancient promises to the Jews. Her life after Jesus grew up is, however, scarcely mentioned in the New Testament. She disappears from history early in the life of the early church in Jerusalem. All else we know about her is legend or tradition, most of which began in the Third Century.

One feminine figure does stand out in the New Testament and that is the true church as the "Bride of Christ." The concept of a consort for God is not new to the New Testament for Yahweh calls himself the husband of unfaithful Israel, his wife, in such Old Testament books as Hosea. (The woman in Revelation 12 is symbolic of the nation Israel as can be seen from the context.)

The Book of Proverbs interestingly personifies Wisdom as a woman present at the creation of the world:

"Does not wisdom call, and understanding raise her voice? On the heights beside the way, in the paths she takes her stand; beside the gates in the front of the town, and the entrance of the portals she cries aloud: 'To you, O men, I call, and my cry is to the sons of man. O simple ones, learn prudence; O foolish men, pay attention. Hear, for I will speak noble things, and from my lips will come what is right; for my mouth will utter truth; wickedness is an abomination to my lips. All the words of my mouth are righteousness; there is nothing twisted or crooked in them. They are all straight to him who understands and right to those who find knowledge. Take my instruction instead of silver, and knowledge rather than choice gold; for wisdom is better than jewels, and all that you may desire cannot be compared with her.

"I, wisdom, dwell in prudence, and I find knowledge and discretion. The fear of the Lord is the hatred of evil. Pride and arrogance and the way of evil and perverted speech I hate. I have counsel and sound wisdom, I have insight, I have strength. By me kings reign, and rulers decree what is just; by me princes rule, and nobles govern the earth. I love those who love me, and those who seek me diligently find me. Riches and honor are with me, enduring wealth and prosperity. My fruit is better than gold, even fine gold, and my yield than choice silver. I walk in the way of righteousness, in the paths of justice endowing with wealth those who love me, and filling their treasuries.

"The Lord created (possessed) me at the beginning of his work, the first of his acts of old. Ages ago I was set up, at the first, before the beginning of the earth. When there were no depths I was brought forth; when there were no springs abounding with water. Before the mountains had been shaped, before the hills, I was brought forth, before he had made the earth with its fields, or the first of the dust of the world.

"When he established the heavens, I was there: when he drew a circle on the face of the deeep, when he made firm the skies above: when he established the fountains of the deep: When he assigned to the sea its limit, so that the waters might not transgress his command, when he marked out the foundations of the earth, then I was beside him, like a master workman; and I was daily his delight, rejoicing before him always, rejoicing in his inhabited world, and delighting in the sons of men.

"And now, my sons, listen to me: happy are those who keep my ways. Hear instruction, and be wise, and do not neglect it. Happy is the man who listens to me, watching daily at my gates, waiting beside my doors. For he who finds me finds life and obtains favor from the Lord; but he who misses me injures himself, all who hate me love death.'" (Proverbs 8:1-36)

## Christ, the Wisdom of God

Some translations of Proverbs 8:22 suggest that God "created" Wisdom as the "first of His acts of old"; however, the Hebrew word used is not really the word for create but a word meaning "to possess." Thus, there is reason to suppose that Wisdom as she is described in Proverbs is an eternal attribute of God, and that She was present and involved in the creation of the universe as well as in its present administration and government.

First Corinthians 1:24 clearly states that "...Christ is the Wisdom of God," so many conservative Bible scholars do not hesitate to equate the woman Wisdom in Proverbs with feminine characteristics of the Godhead. Thus Jesus - the Son of God, the last Adam, a whole man who heads a new race of redeemed men and women - is both the **Logos** (Word) of God and also the **Sophia** (Wisdom) of God, representing in Himself masculinity and femininity in perfect balance.

Another suggestion is to view Wisdom's attributes as a wedding dowry God the Father gives the Church that she might be a Bride fit and endowed for Jesus her Bridegroom. In this case it is the work of the Holy Spirit which is highlighted as for example in the Scripture saying that He, the Spirit, is the "...which is the **guarantee** (the earnest, or downpayment), **of our inheritance** until **we acquire possession of it....**" (Ephesians 1:14) That is, we sinners, who have been called to form the church, the Bride of Christ, have no merits of our own, no natural beauty and no royal

79

qualities that would in any way fit us to be wed to the Lord Jesus, the King of kings. We have no natural right to participate in the "Marriage Supper of the Lamb" unless we are given a suitable wedding dowry (imputed righteousness) by the Holy Spirit. In this view, God the Father has elected to present a beautiful bride, "without spot or blemish" to his beloved Son as a love-gift and He has sent the Holy Spirit out into the world to call her and prepare her. Since wisdom, the proper application of knowledge, is generally learned through experience, it is 2000 years of experience with the faithful strivings of the Spirit of God that help to bring the church to her full glory ready and fit to be presented to the Bridegroom.

The first Eve was taken from the side of Adam; thus, the feminine was inherent and potential in man as he was created. In a similar manner, we can think of the church as the last Eve, taken from the wounded side of Christ on the cross, purchased by His blood, and called to be presented to him as the reward for his supreme sacrifice of self-giving love.

It is speculative to go much further. However, God is certainly complete without man. Before the worlds were made, God was a complete and whole being called "Love." God does not "need" man, nor does all of creation add anything to what God already was and is. Thus, it is safe to assume that the Persons of the Godhead eternally engage in the giving and receiving, and the pouring out of love among themselves, whether or not men, or created things, are there to participate.

Although His feminine side is hidden and not obvious we must assume that God is as much feminine as He is masculine. In keeping with Oriental custom, God evidently hides His feminine attributes modestly behind a veil or an Oriental wooden lattice. To His children, those who are willing to meet Him on His own terms of intimacy, He is as much a tender, healing, nurturing Mother as He is a kind and loving and all-sufficient Father. To the world, He is the Sovereign Lord, Righteous Judge, and Omnipotent Ruler of all.

When the Son gives Himself to do the Father's will, it is because the Son loves the Father and the Father loves the world. When the Father and the Son send the Spirit into the world, it is the Spirit's love which moves Him to accomplish the tasks for which He was sent. God has, for His own good purposes, brought a "fourth party" into that Divine Round of love, which has existed from all eternity. We who form the company of the redeemed have been graciously invited to share with Christ as joint-heirs in His kingdom.

We are to rule with Him as the true Queen of Heaven alongside the King. Although we may not yet see masculine and feminine in balance in our world, and in our finiteness know next to nothing about the complex triunity of God, Psalm 45 shows that masculinity and femininity are in harmony in the Royal Courts of heaven:

"My heart overflows with a goodly theme;
I address my verses to the king;
my tongue is like the pen of a ready scribe.
You are the fairest of the sons of men;
grace is poured upon your lips;
therefore God has blessed you for ever.
Gird your sword upon your thigh, O mighty one,
in your glory and majesty!
In your majesty ride forth victoriously
for the cause of truth and to defend the right;
let your right hand teach you dread deeds!
Your arrows are sharp
in the heart of the king's enemies;
the peoples fall under you.
Your divine throne endures for ever and ever.
Your royal scepter is a scepter of equity;
you love righteousness and hate wickedness.
Therefore God, your God, has anointed you
with the oil of gladness above your fellows;
your robes are all fragrant with myrrh and aloes and cassia.
From ivory palaces stringed instruments make you glad;
daughters of kings are among your ladies of honor;
at your right hand stands the queen in gold of Ophir.
Hear, O daughter, consider, and incline your ear;
forget your people and your father's house;
and the king will desire your beauty.
Since he is your lord, bow to him;
the people of Tyre will sue your favor with gifts,
the richest of the people with all kinds of wealth.
The princess is decked in her chamber
with gold-woven robes;
in many-colored robes she is led to the king,
with her virgin companions, her escort, in her train.

With joy and gladness they are led along
as they enter the palace of the king.
Instead of your fathers shall be your sons;
you will make them princes in all the earth.
I will cause your name to be celebrated to all generations;
therefore the peoples will praise you for ever and ever."

◆

## Notes To Chapter Four

1. For example, see, *The Moment of Creation* by James S. Trefil, (Charles Scribner and Sons; New York, 1983), or Steven Weinberg's, *The First Three Minutes* (Basic Books, Inc.; New York, 1977). When men were learning how to build atomic weapons, enormous computers were required to study the fission and fusion processes in various bomb configurations so that they could be designed properly. Later, these powerful computer resources were directed towards modelling the thermo-nuclear processes that are thought to energize the stars, and also to modelling the Big Bang as a gigantic explosion. The latter must be studied by performing detailed computations covering a range from the smallest imaginable fraction of a second to billions of years.

2. The reader who would like to read a thorough, basic and clear description of the physics of light will surely enjoy, as I have, physicist Michael I. Sobel's book, *Light* (The University of Chicago Press; Chicago, 1987). To a physicist, light is "electromagnetic radiation" ranging from gamma rays, x-rays, ultraviolet, visible, infrared light, through radio waves. The total range spans wavelengths from a thousandth of a billionth of a meter to thousands of meters. The human eye responds to radiation of wavelengths between 0.0004 to 0.0007 centimeters only.

3. Notable among them is Thomas G. Barnes', *Space Medium: The Key to Unified Physics* (Geo/Space Research Foundation; P.O. Box 1350; El Paso, Texas 79913).

4. See for example, Ronald Youngblood, Ed., *The Genesis Debate* (Thomas Nelson Publishers; Nashville, 1986). A number of

excellent Bible commentators on Genesis will be found on the shelves of Christian book stores today. Even reprints of "old classics" can be very helpful to a man or woman of scientific inclination who wishes to better understand the Book of Beginnings. Ray Stedman's books, *The Beginnings* and *Understanding Man* give an outstanding analysis of the first eleven chapters of Genesis, (Word Publishers; Waco, Texas, 1978). An inspiring broad based treatment of theological issues and deeper realities is Erich Sauer's *The King of the Earth: The High Calling of Man according to the Bible and Science* (Ronald L. Hayes Publishers; Palm Springs, CA. 92263; 1981). Henry Morris' *The Genesis Record* (Creation-Life Publishers; San Diego, 1976), includes many insights into Genesis that connect with science in a very helpful way. H. C. Leopold's *Exposition of Genesis* in two volumes is an outstanding older commentary (1942), still in print, (Baker Book House; Grand Rapids, 1984).

5. Robert V. Gentry's life-long and monumental work, described in his book, *Creation's Tiny Mystery* (Earth Science Associates; Box 12067, Knoxville, TN, 37912-0067), is a careful study of "pleochroic halos" in rocks caused by the short-time decay of radioactive atoms embedded in the host rocks. His data clearly shows that *either* ancient rocks of the earth were formed suddenly in a brief period of time, *or* that the radioactive decay of presently unstable atomic nuclei was turned on at a definite date in time after creation. Gentry has been attacked for years by anticreationists, but to no avail. His work still stands as scientifically valid and irrefutable as far as any evidence I have seen to the contrary.

6. C. G. Jung, *Psychology and Religion* (Yale University Press; New York, 1939).

7. See Alexander Hislop, *The Two Babylons* (Loizeaux Brothers, New York).

**Chapter Five**

# The Complexities of Time

## Is The Universe Really Ancient?

Virtually all modern geology and astronomy textbooks today take it for granted that the solar system is at least four or five billion years old, and it is now assumed such great ages are gospel truth. Anthropologists take it for granted that man is at least several millions of years old. Of course, there is *some* apparent evidence to support the case for a very ancient universe.

However, there is other data that can be brought forward to suggest the universe is of recent origin. The reknowned and prolific creation scientist, Dr. Henry Morris, lists sixty-eight global processes indicating recent creation in one of his many books.[1] Long before I became a Christian, but soon after I was educated on the theory of evolution as established fact, I wondered why it was that known history of human civilizations began so recently compared with the assumed billions of years of the history of the universe and the supposed many millions of years of man's upward evolution. The archaeological record of cultural man is almost entirely a record stretching only thousands, not millions of years, into the past. The nature of time and the age of the universe have always fascinated me, and I have been able to follow the arguments of both creationists and evolutionists finding truth on both sides. But when I became a Christian and found in experience that the Old Testament was historically accurate and that dozens of ancient prophecies had already been fulfilled, I wondered how the supposed brief history of man on earth, as revealed in Scripture, could ever fit in with the (recent) scientific theory that the universe was extremely old.

Only in the past 150 years or so has Western science come to believe in a very old universe as opposed to a recent creation. In fact, the record of the development of the theory of evolution, and the modern geological model of very great geological ages is a quite revealing story.[2] Before our modern era, many valid and cogent arguments were advanced for a universe of recent origin; however, today most of these have been dismissed without being refuted. Thus, they are still valid arguments.

## What the Bible Actually Says

Many people today assume that the Bible *directly* teaches a recent creation. Jokes are made about Archbishop Ussher's alleged assignment of the date, day, and hour of creation in 4004 B.C. These derogatory remarks come from the pens of learned men of science who apparently have little familiarity with the Bible and, of course, no personal knowledge of the living God in most cases. The Bible actually opens with the statement **"In the beginning God..."** without making any reference to date and time. In both Hebrew and Greek, the idea of "the beginning" means the "indefinite distant past." This is not to suggest that man's early history fades into obscure mists of mythology as we go backwards in time, but that God has not revealed all that we would like to know about the exact "time" of the creation of all things.

Like Genesis, the Gospel of John opens with the words, **"In the beginning was the Word, and the Word was with God, and the Word was God."** It is assumed in Scripture that God always was, always will be, and is unchanging, and **"Jesus Christ is the same yesterday, today and forever."** (Hebrews 12:8)

As I attempted to show in the preceding chapter, I believe time as we now know it may only have become well-defined after the fall of man. However, if we begin to measure history from the exit of Adam and Eve from the garden of Eden, then the time scale of human history seems indeed to be at most a few thousand years - according to the Biblical record. The Bible is an exceptionally accurate document, and there is no ancient document for which we have better manuscript authority (or evidence), or into which more man-years of intense scholarship have been invested in recovering the original text.

I do not hesitate to state my personal conviction that God is the actual Author of the Book and that it is accurate not only in what it says but in the very ordered arrangement of each letter, in every "jot and tittle." In fact, I am also keenly interested in computer research which analyzes the ordered sequence of letters and the significance of the numerical value of the Hebrew and Greek letters of the text.[3] The Old Testament genealogies have very few gaps in them and are actually quite complete so that one can estimate the time of Adam, the first man, as falling only a few thousand years before Christ.

## Science Can Be Wrong at Times!

The one main reason modern scientists have held to an old universe has been radiometric dating methods. I will introduce evidence in Chapters Six and Seven to show that there is now good reason to believe radioactive decay rates have not been constant but have decreased rapidly with time. This simply means that radioactive clocks have not been running at the same rates as gravity clocks. From this recent analysis of what is essentially scientific data, **I believe the universe can be shown to be young after all!**

Before considering the new evidence for a recent creation, I would like to discuss the nature of time and the dimensions of time as revealed by Scripture. My purpose is to help the reader begin to think Biblically about the world we live in. In my experience, most of us are are so steeped in Western cultural and scientific ways of rational thinking that we have largely lost touch with the ways the Bible communicates truth at several levels.

In Chapters One and Two, I emphasized that the spiritual realm, what the New Testament calls "the heavenly places," is more important realm than the physical. Time exists in both realms, the spiritual and the physical, but does not have the same quality or the same measure in the two realms. Science concentrates on the physical, and a balanced view of things, therefore, requires a bit of theology.

## The Old Testament View of Time

The Hebrew concept of time found in the Old Testament is concerned more with the *quality* of time as it relates to hail, rain, summer, and harvest or to "evil days" or "prosperous times" than to clock or calendar time. Furthermore, the Old Testament teaches by means of stories, historical examples, and case histories of God's dealings with men and angels. It uses poetic images, dreams, visions and providential arrangements of circumstances to indicate God's invisible workings in human affairs from behind the scenes of history.

The Old Testament gives us a record of patriarchs and races, nations and kings. It is a selective record narrowing down to focus on the bloodline leading to the Messiah. Israel is at stage center, all directions are measured from Jerusalem, and the relationship between the Israelites and their God determines their prosperity or adversity in the land (*eretz yisrael*). The historical record of the Old Testament reveals national deterioration and repeated failures by men, but persistent, gracious intervention by God who sovereignly works out His grand strategy down through the ages. Israel typifies God's dealings with the nations. From Israel the Messiah has come, and through Israel will come the ultimate salvation of the nations when Messiah returns.

The Old Testament does not often speak at all about the affairs of other nations unless they impinge on events concerning Israel. Little is said about earthquakes, natural disasters, storms, or cosmic events unless such happenings relate to Israel. The purpose of the record is mostly moral and ethical. Because He is a personal God who makes covenants, Yahweh is evidently much more interested in helping men to know Him and to understand themselves than He is in teaching us science. Concerning the Old Testament, Paul plainly says in First Corinthians, 10:11, that **These things happened to them** (to the Old Testament fathers) **as types, but they were written down for our instruction, upon whom the end of the ages has come."** (My translation)

## Interruptions in Time in the Bible

In discussing the nature of time with a sociologist friend, Dr. Randy Pozos, I was reminded by him of something that I had at that time forgotten (not knowing Hebrew) - that the Hebrew language has no verb tenses in the usual sense familiar to us who speak English. This is consistent with the earlier comment that to the Jewish way of thinking, the quality of an event or happening becomes more important than the minutes or hours (the measure) the event occupies in four dimensions.

For example, in the Old Testament we have mentioned Joshua's "long day," (c.1420 B.C.) when the sun conveniently stood still for about a whole day, so Joshua could finish an important battle against the Amorites. (The battle is described in Joshua Chapter 10.) The LORD also conveniently arranged an exceptionally heavy hailstorm at the same time, suggesting that something radical happened to the earth's normal weather patterns at the same time. Of course, what actually took place in nature would be, to us, of enormous scientific importance to learn more about. However, the Bible makes the stopping of the earth's rotation on its axis and the fall of enormous, deadly hailstones *incidental* to the main purpose of the narrative which was recorded to show how God can use supernatural means to deliver His people. I have no doubt that what occurred was a rare-but-real intervention of God into the normal 24-hour day.

Some day I hope to read about supporting evidence for an unusual event such as a large meteor striking the earth or a great volcanic explosion that would correlate con-clusively with Joshua's Long Day. The idea that God should interrupt the normal flow of time for a moral reason may strike us as "unreasonable," and, of course, explaining how He does it, (the laws of physics being what they are), is something I have never seen anyone successfully explain. Some have felt the earth would fly apart if its rotation were stopped. This assumes that God lacks sufficient power to coordinate and control all related forces such as tides and stresses in the crust.

About 714 B. C. King Hezekiah faced the crisis of early death and asked God for help, (II Kings 20). He granted the king fifteen more years. As a sign, God caused the sun dial in the palace to move backwards "ten steps." Perhaps the reversed motion of the sun dial was caused by some sort of wobble in the earth's rotation.[4]

The Hebrew idea of continuous present tense is found in the covenant name of God (one of many names for God in the O.T.). This is the God who revealed Himself to Moses at the burning bush in Sinai saying, **"I am Who I am. Tell Pharaoh, 'I AM' has sent you."** This could be translated equally well as **"I Will be Who I Will Be."** The name **YHWH** (**Yahweh**, or **Jehovah**) is simply derived from the verb "to be." God is the great "I AM" in the sense that each of us is a little "i am." In reading the Gospel of John it is helpful to note that Jesus used the term "I am" a number of times in the sense of the meaning of Yahweh. For instance He said, **"...before Abraham was, I am."** (John 8:58) Jesus was much more aware of the eternal dimension than we are. He dwelt in eternity in some sense the whole time he was present on earth as the Man Christ Jesus. Thus, some of the accomplishments by Jesus at points in time while He was on earth sent ripples into eternity which changed both the past and the future! As God is eternal and outside of time, so our human spirits are also eternal. However, our bodies are fallen, subject to death, and not yet redeemed. It is the fact that our spirits live in *bodies* that places us in contact with the physical world and limits our experience of time.

## Subjective Time

It is important to bear in mind that various dimensions of time are known to man. First, there is subjective time, which is the appearance of time to our sense of consciousness. It cannot be measured, but varies from what we perceive as happening in a flash to dragging on "forever" - while the clock on the wall may tick off only minutes.

Many of us remember how time appeared to move very slowly during childhood. A single summer day seemed to last forever, and the interval between Christmases and school vacations was an eternity. Later in life, some of us look back and see that decades have passed almost as if they were but months. In sudden accidents, some have reported that their whole lives flashed before their eyes in great detail, in a few seconds. When we dream at night, what seems to be many hours of time is shown by REM (Rapid Eye Movement) sleep patterns to be only minutes of elapsed time.

Biological time has to do with wildlife migratory patterns, animal hibernation, biorhythms, jet lag, circadian (24-hour) patterns and menstrual cycles - numerous phenomena in nature that are loosely coupled to dynamical time (that is, to months and seasons). Although such biological time clocks are mysterious and poorly understood, they are probably closer to the way God keeps time, if we remember that the Jewish calender is based on the lunar month, the cycle of harvest, and the motion of the earth, moon, planets, and stars. Seen in this light, the scientist's way of keeping time - with precision quartz clocks and atomic resonators - is arbitrary and less "absolute" than God's heavenly clocks and calendars.

# Linear Time and Cyclical Time

The Hebrew view of time also includes the concept that time moves from event to event in a line - not a straight line, to be sure, but towards a goal. The goal is always the future, yet to be fulfilled in history. Bible prophecies frequently have both an immediate and a long-term fulfillment, for example. Sins have consequences, moral choices lead to measurable results for good or for ill, and history proceeds towards the definite outworking purposes of God.

A consummation of the ages lies ahead, for which all else has been but a shadowy preparation. I mention this because in both ancient Greek culture, (among the Pythagoreans, Stoics and Neoplatonists), and in Hindu culture (especially during the Vedic period, 1500-600 B.C.), one runs onto the concept of circular, or cyclical time. This is sometimes symbolized by the *uroboros*, the snake chasing his own tail. In this view of time, the beginning leads back around to the end, and the cycle starts all over again. The Babylonians, ancient Chinese, Aztecs, Mayans, and the Norse had cyclical calendars. In pantheistic religious systems of thought, the universe is often depicted as going through great long epochs of rebirth, growth, decay, and destruction. The Hindu cycles, for example, range from 360 human years, to 300 trillion years (which is the lifetime of the gods before their rebirth.) Reincarnation (which has no basis in the Bible at all), springs from such an Eastern pantheistic point of view. Augustine was among the first to insist on linear time as opposed to cyclical, since he observed that many important events in the Bible clearly happened one time only. I suppose, since clocks were not well-developed until the 14th Century, it was easier to imagine events in history as recurring since the four seasons and patterns of the stars in the heavens were cyclical.

The Bible depicts the human race as having a definite clear beginning, a history which has been accurately recorded by God, and an approaching day of judgment when all men will be evaluated justly by their Creator. The fact that "books are to be opened" on judgment day means God keeps track of detail even if we do not, and that He pays attention even to the numbering of the hairs on our heads. He will see to it that truth and justice ultimately prevail no matter how grim things seem to us at the moment!

# Time in the New Testament

The New Testament appeals to reason, to the conscience, and to the rational mind, to communicate the same truths that are found in the Old Testament in story form. Someone has said that the Old Testament appeals to the right side of the brain and the New, to the left side. The message is really the same, but it is presented in two differing formats in the two halves of Scripture. The New Testament message is addressed not only to the Jews but to the pagans, the Goyim, the non-Jewish world. When the New Testament was written down in the First Century, A.D., Greek and Roman culture and government dominated much of the ancient world.

The Greek language of the New Testament refers to time as measured in *"chronos"* and *"kairos"* - "times and seasons." The meanings of the Greek New Testament words for times and seasons add more to an understanding of the complex nature of time in our universe. *Chronos* means *quantity* of time, space of time, duration, succession of moments, length of time, or a bounded period of time. To understand this word, we should read the passages of the New Testament where *chronos* is used. These include Matthew 2:7, Luke 4:5; 8:27; 20:9, Acts 20:18, Romans 16:25, and Mark 2:19. *Kairos* refers to the *quality* of time or season, the epoch characterized by certain events, the decisive quality of happening, an opportune time, or a fortuitous moment.

The gifted Archbishop Trench says, **"The 'seasons' are the critical epoch-making periods foreordained of God, when all that has been slowly, and often without observation, ripening through long ages is mature and comes to birth in grand decisive events, which constitute at once the close of one period and the commencement of another. Such, for example, was the passing away of the old Jewish dispensation; such, again, the recognition of Christianity as the religion of the Roman Empire; such the conversion of those outside; such the great revival which went along with the first institution of the Mendicant Orders; such, by still better right, the Reformation; such, above all others, the second coming of the Lord in glory."**

*Kairos* is used in such passages as Romans 5:6, Galatians 6:10, Matthew 13:34, 26:18, Revelation 12:12, I Peter 1:11, and Luke 4:13. In the New Testament we have expressions like **"times of** refreshing" (Acts 3:19), "**times of ignorance**" (Acts 17:30), and "**the times of the Gentiles**" (Luke 21:24).

## Times of Stress

To give an example, the expression "**times of stress**" occurs in one of the most interesting passages in the New Testament, Paul's second letter to Timothy 3:1-5. Our understanding of the message is enriched by looking up the individual Greek words in this passage in a lexicon.[5] The passage in question reads as follows:

"**But understand this; that in the last times there will come times of stress. For men will be lovers of self, lovers of money, proud, arrogant, abusive, disobedient to parents, ungrateful, unholy, inhuman, implacable, slanderers, profligates, fierce, haters of good, treacherous, reckless, swollen with conceit, lovers of pleasure rather than lovers of God, holding the form of religion but denying the power of it. Avoid such people.**"

The entire New Testament uses the term "last days" to refer to the entire 2000 year interval between the first and second advents of Christ. Christ was born "late" in history as God measures time. He will return after recurring cycles of stress have plagued mankind. These cycles will come with repeated frequency and intensity as the age draws to a close. They will also be less and less local and more and more global. For example, only in our century have we had "World" Wars. The present world economy is another example. A recession in one nation these days affects the world economy creating a crisis not easily corrected by any individual sovereign nation.

It is not possible for us to anticipate where and when the next "time of stress" will befall us, nor can we tell what form it will take. Thus, we cannot plan ahead very well, so we must take one day at a time as Jesus advised us in the Sermon on the Mount, "**...Sufficient unto the day is the evil thereof.**" (Matthew 6:34) (KJV) During these times of stress, the real character of human beings surfaces, raw, ugly sores open in society, and the situation becomes dangerous and violent. Astrologers explain that such times are at least partially caused by "unfortunate" aspects and alignments of the planets.

93

# The Mysterious Flow of Time

Although time is measured in history in terms of clocks and calendars, it is also articulated into seasons. They bend, and stretch, and unfold as God periodically moves the course of history in a different direction. Or, at the last minute, He postpones the consummation of events that seem to be right upon us at a time of crisis.

Twice the Bible makes important statements (consistent with each other) that suggest the fundamental nature of time, and many aspects of the actual course of history, presently escape our understanding to a large degree. That is, some of the essentials of time are hidden from us by God. Solomon writes, **"I have seen the business that God has given to the sons of men to be busy with. He has made everything beautiful in its time; also he has put eternity into man's mind, yet so that he cannot find out what God has done from the beginning to the end."** (Ecclesiastes 3:10-11)

Just as we cannot figure out God's ways and understand precisely how He works, so also we may not notice events that are actually crucial to His plans and programs. And we may mislabel other events in history as important when actually they turn out to be unimportant in the long run. Most of Israel totally missed the many prophetic fulfillments that took place during the First Advent of their Messiah, Y'shua (Jesus), for instance. Only afterwards did His followers figure out what actually had been happening in God's plan as revealed in the Old Testament.

Similarly, our knowledge of what actually happened in the past is inadequate; the details of what was important and what was not are obscured in the mists of time. It is most difficult for historians to re-construct what actually happened. Likewise, we cannot predict what events will unfold tomorrow with any certainty nor set a date for the return of Christ. Yet we are restlessly preoccupied with time and frustrated when we cannot unravel its secrets with all of the precision a modern atomic clock can give us!

When Jesus left His disciples forty days after His resurrection, ascending into the cloud, (that is, the Shekinah Glory) from the Mount of Olives in Jerusalem, His disciples were anxious for word of His return. Jesus told them, **"...It is not for you to know times or seasons which the Father has fixed by his own authority."** (Acts 1:7; compare with Matthew 24:36). This, of course, means

94

that all attempts to set dates for the Last World War and the second coming of Jesus are wasted effort. The ages past are also difficult for us to unravel and must remain full of mystery. The tapestry of the past has many folds, and we easily lose track of most of them in our feeble attempts to trace history backwards.

I mentioned earlier in this book that God's actions in eternity can affect past, present, and future (as experienced by mankind), simultaneously. A certain action of God completed in the past can have on-going and lasting results. Other activities of God, such as His expressions and grace and mercy towards us all, continue day after day. Certain events, such as the "appointed" hour we die or the Day of Judgment, are fixed in the future, predetermined by God. Since God is more concerned with the quality of time than the quantity or measure of time, we can all expect to experience time differently in eternity depending on the quality of our lives during our present training on earth.

## An Empty Universe without Man Makes no Sense

I believe the universe was created for man. An empty universe inhabited only by God and the angels makes little sense to me. God does not "need" a universe, nor does he need man to add to His Being. He is fully sufficient and complete in all His attributes so our creation adds nothing to His essential nature. The universe was made as a home for man, and man was made for fellowship with God. Of course, God pronounced nature "good" and valuable to Him before He placed man on earth. I do not believe (intuitively) that the universe sat empty for aeons before man arrived on the scene. God proceeded to create the universe step by step in an orderly way, and when He had it prepared, He made man and placed him squarely in the center of things to understand and to rule over what had been created.

The notion of an originally upright, unflawed universe suggests that the moon and planets may have once been more beautiful, more pristine, and more "inhabitable" than they are now. I tend to feel that some sort of cosmic disaster has already occurred and that there is ample evidence now of destructive forces at work that were not put here by God. The Biblical view also contradicts the notion that man is improving and society is advancing morally and socially. Rather, it is the grace of God which makes life bearable and prevents mankind from self-destruction.

The original creation was "good" (unmarred, flawless). Then the angels fell and later man fell, placing a "curse" on the physical world which has not yet been lifted, as will be discussed in Chapter Eight. The fall of man and the fall of Satan have, I believe, made fundamental changes in certain laws of physics and biology as well. And I believe that the nature of subjective time has changed since creation. Also, man in his present condition is constrained to a rather limited "one-dimensional" time frame whereas before the fall, our first parents enjoyed a multi-dimensional quality of time much richer than we can even begin to imagine.

## A Glimpse into Eternity

I believe we can show that the Biblical view of time in heaven found in the New Testament is actually multi-dimensional. For example, in the Book of the Revelation we see scenes taking place on earth in human history and scenes in the heavenly places going on at the same time. Time in heaven moves in the forward direction as it does on earth, for example Revelation 8:1 describes a period of silence in heaven lasting "about half an hour". But time in heaven has a quality and a pace different from time on earth. A good example of an event occurring in "eternity" is found in the Gospels: one day Jesus stepped up to the top of Mount Mizar above Banias (ancient Caesarea Philippi) in northern Israel and was transfigured before His frightened disciples, Peter, James, and John. Appearing with Him were Moses and Elijah, alive and well. They all conversed together as if contemporaries of one another, oblivious to the years that had separated them - by our way of reckoning time. This incident (recorded in Luke 9:28-36; Matthew 17:1-8; and Mark 9:2-8) shows that all the usual rules and constraints of time (as we commonly think of them) were momentarily lifted. Thus, it was not only possible for men from ancient times to appear alive in the presence of the disciples of Jesus, but also for Jesus to assume His glorified body all at the same "time." Luke's account is as follows:

**"And he said to all, 'If any man would come after me, let him deny himself and take up his cross daily and follow me. For**

whoever would save his life will lose it; and whoever loses his life for my sake, he will save it. For what does it profit a man if he gains the whole world and loses or forfeits himself? For whoever is ashamed of me and of my words, of him will the Son of man be ashamed when he comes in his glory and the glory of the Father and of the holy angels. But I tell you truly, there are some standing here who will not taste death before they see the kingdom of God.'

'Now about eight days after these sayings he took with him Peter and John and James, and went up on the mountain to pray. And as he was praying, the appearance of his countenance was altered, and his raiment became dazzling white. And behold, two men talked with him, Moses and Elijah, who appeared in glory and spoke of his departure, which he was to accomplish at Jerusalem. Now Peter and those who were with him were heavy with sleep, and when they wakened they saw his glory and the two men who stood with him. And as the men were parting from him, Peter said to Jesus, 'Master, it is well that we are here; let us make three booths, one for you and one for Moses and one for Elijah' - not knowing what he said. As he said this, a cloud came and overshadowed them; and they were afraid as they entered the cloud. And a voice came out of the cloud, saying, 'This is my Son, my Chosen; listen to him!' And when the voice had spoken, Jesus was found alone. And they kept silence and told no one in those days anything of what they had seen." (Luke 9:23-36)

Perhaps another example of the dimensions of time and eternity will be helpful to the reader. Consider the various time frames that are involved in the writing, printing, and reading of a book, for example, a mystery novel. Perhaps the author took a year to write the manuscript, but drew from many years of personal experience and from his own reading of history. Suppose six months elapse before the book is on the market and reaches the reader. The reader then begins the book, and after a period time of intermittent reading, finishes it. (The reader can even skip ahead to the end, if he wishes, to see how it all turns out). Internal to the book is the time frame of the story, which may include flashbacks in the lives of some of the characters. After reading the book, it goes on the library shelf, but the reader retains a summary version, condensed in his memory. He is free to recall the book, or read it again. In this example one can count half a dozen, or more, different time frames all co-existing!

## Time in a Resurrection Body

After His resurrection, Jesus further demonstrated the capacities of His resurrection body by appearing and disappearing at will among His disciples, in the days between the resurrection and the ascension. From such records in the Gospels, we can conclude that resurrection bodies are equipped for multi-dimensional space and time travel. Jesus ate food and could be touched and felt, in His resurrection body. He did not return in a ghost-like, shadowy form. Further discussion of the nature of the resurrection body has been written by the late Canadian scholar Dr. Arthur Custance.[6] In his two letters to the Corinthians, the Apostle Paul clarifies the nature of the resurrection as the point a believer steps out of this time frame of human history bypassing intermediate (future) times to arrive at the resurrection at the exact same instant all other believers do. In the experience of the Christian, one's personal death corresponds exactly with the Second Coming of Christ, though this event will also happen on earth at the definite date and time in recorded human history. This is what Paul meant when he said to be absent from the body was to be at home with the Lord, not as a spirit, but in a resurrection body along with everyone else who knows God.

## A Personal Eschatalogical Point of View

For the benefit of the reader who is interested in my own view of eschatology (the doctrine of the "last things" at the close of the present age), I believe in what is known as the pretribulation rapture of the church. I am also a premillenialist, believing in a literal reign of Christ on earth for one thousand years before the "new heavens and the new earth" are finally brought in. I believe Christ will return to earth at a definite point in the future to raise the believing dead of the Christian era. All living Christians then on earth will be caught up with the dead who have been raised, as described in I Thessalonians 4. This will unify and unite the Bride of Christ as a completed body of all those who have believed in Jesus since Pentecost. In the spiritual realm with Christ, all these believers will experience the judgment seat of Christ and then the so-called "Marriage Supper of the Lamb." These latter events run on the time scale of heaven's clock rather than on the clock of earth's dynamical time.

This return of Christ for His church will be more or less silent and invisible, creating but a momentary, passing flurry of concern on earth. Jesus will actually remain on earth during the ensuing seven-year period, with His church, but unseen and invisible to mankind. His activity will cause local stirs of interest as he recruits, dispatches, and leads the 144,000 converted Jews of that era so that they undertake vigorous world-wide evangelism, bringing many thousands - if not millions - of "tribulation saints" into the kingdom of God. These saints will be a body of believers separate from both believing Israel and the church.

The seven-year tribulation period on earth, marked by the sudden appearance of the "man of sin" and his Jewish cohort, the false Messiah of Israel, will be divided into a three and a half year period of apparent world peace, and a second three and a half year time of terrible destruction on earth. The apparent peace is to be interrupted suddenly by the desecration of the Third Temple in Jerusalem. That event will mark the beginning of the "Great Tribulation," also known as the Seventieth Week of Daniel and the "Time of Jacob's Trouble" spoken of by Jeremiah the prophet. During this time, terrible world-wide judgments, described by the Old Testament prophets, by Jesus in the Olivet Discourse, and by John in the Book of the Revelation, will be unloosed on earth.

Invasion of Israel will follow, threatening the total destruction of that nation, but Jesus Christ will then appear in power and glory with His church and His angels to establish His thousand-year reign on earth. This will be the era when God will make a new covenant with the nation of Israel. That people, Israel, will become the chief among the nations, and God will honor His permanent promises to Israel given to Abraham, Moses, David, and Jesus.

I am a strong supporter of the nation of Israel, recognizing, however, that the believing remnant in that country today is at most a few thousand persons. However, in light of the Old Testament and especially Romans 9-11, I cannot believe that God has set aside Israel forever but will restore them again to Himself. It should be obvious from the book of Zechariah that Christ's return will be to the Mount of Olives, Jerusalem, rather than to Zion, Illinois! Seen from the vantage point of heaven, these events will not occur in the linear time sequence that I have described them. The First Chapter of II Thessalonians, for instance, gives a view of the second coming of Christ as seen from a vantage point in heaven.[7]

## Three Greek Words

Not all Christians will agree with my own personal statement of belief concerning the future. Legitimate differences exist between Christians on many points of Scripture. Since I have attempted to compare and contrast events occurring in heaven with events taking place on earth, perhaps a word of explanation is in order about the future event known as the "Second Coming of Christ." Three different Greek words are used to describe Christ's return. The first word is **parousia** from **para** meaning "alongside" and **ousia** meaning "being" or "presence." Hence, this word means coming alongside and remaining with someone. This word is used to describe the return of Christ silently and invisibly for his church. A second Greek word is **ephiphaneia** meaning "a shining forth". In II Thessalonians 2:8 both of these words occur together, to illustrate the open appearance of Jesus before men following a period of time when He has been with His followers on earth, but unseen by the masses. Finally, the Greek word **apokalupsis** means "to unveil" or "uncover." Seen from the vantage point of eternity, the Second Coming of Christ is one event. For observers on earth, Jesus arrives and remains, for a seven-year period, after which He makes Himself known openly and in glory for the whole world to see. For those who do not read Greek, (and I am one of them), an interlinear Bible with Greek transliterations, and a good Bible dictionary or lexicon, will help enrich one's understanding of the many details contained in the original langauges that are easily lost in our English translations.

## Eternal Life Starts *Now!*

Of course, "eternal life" - which is the free gift of God to all those who receive Jesus Christ as Lord - is a kind of time dimension characterized *not only* by endless duration, but by very high quality. God's time has richness, variety, freedom from boredom and endless diversity. Living in fellowship with Him who is Life is not only liberating but exciting beyond the power of language to describe. Eternity does not mean "timelessness," except perhaps for those in hell!

**"O the depth of the riches and wisdom and knowledge of God! How unsearchable are his judgments and how inscrutable his ways! 'For who has known the mind of the Lord, or who has been his counselor?' 'Or who has given a gift to him that he might be repaid?' For from him and through him and to him are all things. To him be glory for ever. Amen."** (Romans 11:33-36)

## Notes to Chapter Five

1. Henry M. Morris, *The Biblical Basis for Modern Science*, Appendix 6 (Baker Book House; Grand Rapids, 1984).

2. See Ian T. Taylor's *In the Minds of Men: Darwin and the New World Order* (TFE Publishing; Box 5015; Toronto, Ontario, Canada, M4Y 2T1, 1984). Taylor's work is an outstanding account of the history of these two areas of modern science. The author, a research metallurgist by profession, gives a number of excellent examples of scientific reasons men why believed the universe was of recent origin before the theory of evolution and modern geological theory came into vogue. I highly recommend this book, now in its second edition.

3. The Meru Foundation, headed by Mr. Stan Tenen (7 Hillcrest Court; San Anselmo, CA 94960) is doing the most interesting analysis of the underlying structure of the text of the Bible that I am aware of. Tenen's work is not merely numerical analysis or Kabbalistic mysticism, but an attempt to show that there are hidden patterns in the text itself that can be used to derive the geometrical properties of our universe. In recent years, computer research groups in Israel have shown from textual analysis that the Pentateuch was written by Moses, and earlier computer studies in the New Testament have helped to confirm the authorship of a number of New Testament books.

To establish an accurate calendar of events, the Chronology-History Research Institute (P.O. Box 3043; Spencer, Iowa, 51301) is undertaking computer dating of the Bible. This group has issued several important books and publications and has a helpful newsletter, *It's About Time*.

A fascinating compilation of more than forty calendars, ancient and modern, is found in Frank Parise, ed., *The Book of Calendars* (Facts on File; New York, 1982).

4. Among contemporary creationists, Donald E. Patten has written a number of provocative books and articles on catastrophic happenings in ancient times. See his *The Biblical Flood and the Ice Epoch* (Pacific Meridien Press; Seattle, 1966); *The Long Day of Joshua and Six Other Catastrophes* (ibid., 1973); and Six Volumes of which he is the editor, *A Symposium on Creation* (ibid., 1977).

5. A Greek lexicon adds further content to our understanding of this passage:

lovers of self: *philautoi* means self-centered, selfish, a me-first attitude, putting one's own interests ahead of the common good.

lovers of money: *philarguroi* means love of power, prestige, wealth, and greed for worldly success and advancement in society. (While it is commonly said that the Bible teaches that money is the root of all evil, what the Bible actually says is **"the love of money is a root of all evil"**).

proud: *huperephanoi* literally means "to show oneself up"; that is, to have contempt for everyone but oneself, to be aloof, disdainful, and uninvolved.

arrogant: *alazones* means "wandering about," vainglory. It is used to describe the medicine show quack, the man who makes many promises but never delivers.

evil-speakers: *blasphemoi* is, of course, the root of our common English word. It means to insult God and man, especially to tear down the character of another person.

disobedient to parents: *apeitheis* means unwilling to be persuaded by one's parents. It implies loss of respect for authority and an obstinacy that spurns belief.

ungrateful: *acharistoi* literally means "without grace" which includes "grace-lessness" and thanklessness towards God and man.

unholy: *anosioi* means having no decency, debased, decadent, profane; a person who readily indulges in shameless self-gratification. It implies a pollutedness and a lack of righteousness.

inhuman: *astorgoi* is "without natural affection," used to describe the parent-child relationship especially, lack of natural kindness and concern towards others.

implacable: *aspondoi* is "unwilling to sign a truce." This refers to a reluctance on our part to enter into binding agreements such as marriages, jobs, treaties or debts; in other words, irresponsibility.

slanderers: *diaboloi* is a familiar root and refers to the person who enjoys story-telling, gossiping, fault-finding or making innuendos that tear down the character of another.

incontinent: *akrates* is lack of control over one's desires. It implies powerless, impotent, and unrestrained living.

fierce: *anemeroi* ("without gentleness") means savage, without any sympathy for others, brutal, bestial.

haters of good: *aphilagathoi* means "not loving good things." This means associating with base, sleazy people and situations, reading cheap literature and feeding on the degraded things of this world.

treacherous: *prosdotai* means a person who is revengeful, who pays back old grudges and seeks constantly to settle the score with others.

reckless: *propeteis* refers to recklessness, impulsiveness; an unwillingness to wait for wisdom or to think things through; headstrong.

puffed up: *tetuphomenoi* means a person who looks down on others, is conceited and has an inflated sense of self-importance.

6. See his *Doorway Papers* (Zondervan Press; Grand Rapids, 1976) and his books *Journey out of Time* and *Two Men Called Adam*, published by the author. Biblical notions of Time and Eternity are eloquently discussed by C. S. Lewis in his *Mere Christianity* (Macmillan Publishing; New York, 1960); and by Ray C. Stedman in *Authentic Christianity* (Multnomah Press; Portland, 1975). The latter book is unexcelled in helping Christians separate authentic Christianity from the prevailing counterfeit versions common in our time.

7. The most helpful single book on this subject I know of is Ray Stedman's *What's This World Coming To* (Regal Books; Ventura, CA., 1970). Also, Dr. Stedman has a most helpful tape series, *Understanding the Times*, available from Discovery Tapes; 3505 Middlefield Road; Palo Alto, CA. 94306.

# Chapter Six

## Is Light Slowing Down?

## A Revolution in Recent Day Physics

That a major revolution in nuclear physics, astronomy, and cosmology is underway these days is perhaps not obvious to the general public, or even perhaps to the average research scientist who is not working directly in one of these fields. It was but 300 years ago this year that Sir Isaac Newton published his *Principia*, launching the Western world into the era of present-day physics. An explosive increase in the body of knowledge known as "modern physics" has resulted. The most rapid changes in this body of knowledge, however, seem to have occurred in the past few years and appear to be taking place even now at an accelerated rate.

As startling and profound as Albert Einstein's Special and General Theories of Relativity were when they first appeared, shortly after the turn of this century, advances in particle physics and in astronomy in the past three or four decades have been even more radical in their implications. On technical book shelves recently, several books have appeared noting that it may be no mere coincidence that the human species lies more or less midway between the realm of the very small and the astronomically large.[1]

It is now known that certain atomic constants governing the atom and its inner workings are the very same constants that likewise describe phenomena in space-time on the largest scale of observables in the universe. Thus, for some as-yet-unexplained reasons, the realm of the smallest physical observables is coupled to the grandest scale of sizes and events among the stars and galaxies. Modern cosmologists observe not only that the large and the small are interlinked, but also that had there been (at the time the universe began) tiny, insignificant variations in some of these physical constants the end result (today) would have been so radically different that man, the observer, would not be here to see it all, or to write theories to explain how it all happened! I hasten to add that the above remarks come from secular scientists, not from Bible scholars or creationists.

## When World Views Change

All science rests on philosophical or religious presuppositions and basic assumptions made at the start of an hypothesis, as we have noted earlier. Good science means questioning basic assumptions from time to time or altering one's *weltaunschaung* (world-view) in the light of new findings. Today's scientific theories are built on the foundations laid by the previous generation, and a good many of our theories are certainly valid because they work so well and have stood the test of time. But old theories do give way to new, and hopefully a net gain in understanding follows.

Often a new scientific theory is found to fit the experimental data very well - at first - and everyone rejoices. Then more precise measurements are made. When the new data are in, small differences between theory and experiment are frequently discovered. In physics, the numerical discrepancies may, at first, be so small that they are discovered only by taking utmost care to assure accuracy in the measurements and analysis of data. Whenever real discrepancies remain, concerted efforts (often by many research groups) are launched spontaneously to find the reasons for the discrepancies and to revise the older theory.

Growth in science also depends on new ways of looking at old data, at carefully looking for the exceptions to the rule, or by following hunches or intuition or "leaps of faith" to see where they lead. Once in awhile, a truly major upheaval in science occurs. When fundamental notions of science or philosophy change, what has happened is nowadays known as a "paradigm shift." Eventually, social values, customs, and lifestyles change as a result.

The writer of the letter to the Hebrews in the New Testament talks about the course of the age in which we now live, telling us that we can expect radical, sudden, even frightening, changes in our perceptions of things as the age draws to a close. These will come as a result of the breaking through of the eternal and the spiritual into the realm of the physical:

"See that you do not refuse him who is speaking. For if they (the Jews) did not escape when they refused him (Moses) who warned them on earth, much less shall we escape if we reject him who warns from heaven. His voice then shook the earth (at Mt. Sinai); but now he has promised, 'Yet once more I will shake not only the earth but also the heaven.' This phrase, 'Yet once more,' indicates the removal of what has been shaken, as of what has been made, in order that what cannot be shaken may remain. Therefore let us be grateful for receiving a kingdom that cannot be shaken, and thus let us offer to God acceptable worship, with reverence and awe; for our God is a consuming fire." (Hebrews 12:25-29)

Choosing to study observational anomalies (apparent discrepencies in physical laws) that apparently run counter to the prevailing assumptions of the day is not guaranteed to prove popular with all scientists. Many scientists have never taken a class in the history of science so as to be aware of how the body of scientific evidence has developed over time, or they would be, perhaps, less afraid of change. Some researchers may be so engrossed in the excitement of their current studies that they fail to take into account new evidence from other disciplines or to question the assumptions upon which prevailing models rest. Yet many of today's scientific orthodoxies originated from yesterday's unpopular heresies.

## The New "Heresy" of Changing Velocity of Light

I first became aware of the research investigations of Trevor Norman and Barry Setterfield four years ago. I had stumbled across, almost by accident, a technical paper in which they described an analysis of the known experimental measurements to date of the velocity of light. Their data seemed to show that a small (but statistically significant) decrease in "c" had occurred during the past 400 years. I followed the subsequent printed responses solicited from scientists around the world on the issues raised by the original paper and found Norman and Setterfield had competently answered most of the questions raised by critics of their theory. I knew from experience that major changes in scientific theories often begin just this way, and frequently meet with violent opposition from the scientific and religious establishments. I have learned to sort out such new ideas when they appear in print and to pay close attention to a few of these, for often change and progress in science take place from small beginnings.

At first I was interested in the research paper by Setterfield and Norman but cautious and skeptical. I remembered speculations concerning the red shift of light from distant galaxies when I was an undergraduate in physics at San Diego State University (near the famous 200-inch Hale Telescope on Mt. Palomar). The cause of the red shift has since been explained as a result of the expansion of the universe outwards from a point of singularity. These revolutionary ideas in astronomy were not, I recalled, well received by all when they were first propounded in the 1920's and '30's. I had heard of the possibility of "tired light," but always assumed the speed of light had been dependably constant for billions of years.

So out of curiosity I wrote Mr. Setterfield after reading their monograph. I soon received a courteous reply. There followed an exchange of comments, reports, and news articles, and numerous telephone conferences between us. I have since talked to a number of other scientists in the United States and abroad who also take Norman's and Setterfield's work seriously. Last year Mr. Norman was instrumental in establishing an electronic mail connection between our two organizations to facilitate discussions among the three of us.

In all honesty I can say that it has taken me four years to get comfortable (and enthusiastic about) their findings. It has been very good for me to do my homework in the process of evaluating what they have written. I have had to dig out my quantum mechanics, atomic physics, relativity and cosmology textbooks from graduate school at Stanford University and get up to date a bit by reading more recent works. When I learned recently that Norman and Setterfield had carried their work to the stage where a new report had been drafted, I offered my assistance in hopes their findings could be made better known. The result was the publication of a more up-to-date technical, scientific paper which is now available.[2] Since I consider their report an important one, I have included the authors' abstract:

**Abstract:**

"The behavior of the atomic constants and the velocity of light, c, indicate that atomic phenomena, though constant when measured in atomic time, are subject to variation in dynamical time. Electromagnetic and gravitational processes govern atomic and dynamical time respectively. If conservation laws hold, many atomic constants are linked with c. Any change in c affects the atom. For example, electron orbital speeds are proportional to c, meaning that atomic time intervals are proportional to $1/c$. Consequently, the time dependent constants are affected. Therefore Planck's constant, h, may be predicted to vary in proportion to $1/c$ as should the half-lives of the radioactive elements. Conversely, the gyromagnetic ratio, $\gamma$, should be proportional to c. And variation in c, macroscopically, therefore reflects in the microcosm of the atom. A systematic, non-linear decay trend is revealed by 163 measurements of c in dynamical time by 16 methods over 300 years. Confirmatory trends also appear in 475 measurements of 11 other atomic quantities by 25 methods in dynamical time. Analysis of the most accurate atomic data reveals that the trend has a consistent magnitude in all quantities. Lunar orbital decay data indicate continuing c decay with slowing atomic clocks. A decay in c also manifests as a red-shift of light from distant galaxies. These variations have thus been recorded at three different levels of measurements: the microscopic world of the atom, the intermediate level of c measurements, and finally on an astronomical scale. Observationally, this implies that the two clocks measuring cosmic time are running at different rates. Relativity can be shown to be compatible with these results. In addition, gravitational phenomena are demonstrated invariant with changes in c and the atom. Observational evidence also demands consistent atomic behavior universally at any given time, t. This requires the permeability and metric properties of free space to be changing. In relativity, these attributes are governed by the action of the cosmological constant, $\Lambda$, proportional to $c^2$, whose behavior can be shown to follow an exponentially damped form like $\Lambda = a + e^{kt} (b + dt)$. This is verified by the c data curve fit.

(Note: A dynamical second is defined as $1/31,556,925.9747$ of the earth's orbital period and was standard until 1967. Atomic time is defined in terms of one revolution of an electron in the ground state orbit of the hydrogen atom. The atomic standard by the caesium clock is accurate to limits of $\pm 8 \times 10^{-14}$)."

Statistician Alan Montgomery wisely cautions concerning drawing scientific conclusions from statistical analyses: "(1) Statistical Inference is just that, an inference; no proof is implied - just a likelihood; (2) Methodologies which do not have the same systematic error *may* cause the statistic to be invalid; (3) Data which deviate inordinately from other data points can have an undesirable effect on the statistic; (4) The statistic cannot be used unless the time sequence is known; and (5) The error margin is much smaller than the drop in the observed value."

With these qualifications in mind, he says, in regard to Setterfield's early, preliminary (1983) data on c-decay, "It is beyond coincidence in my opinion that all these numbers should show definite trends, in the appropriate direction and in the appropriate amounts. There is no data for a rational mind to reject Mr. Setterfield's conclusion of c decay and every statistic leads the reasonable mind to accept it."[3]

Surprisingly to me, some of the strongest criticism challenging Setterfield's and Norman's work has come from several creationists who apparently feel this new theory threatens the time-honored turf they claim as their own! The "not-invented-here" syndrome perhaps is to be found among some of the Christians who are supposed to know more about creation than the rest of us! Much of the material contained in the Australian report is advanced physics, to say the least, and in spite of the apparent accuracy of the main premises, some of the details may need "fine-tuning." This is not unusual when a major new work of science first becomes accepted, even by a small group of scientists.[4]

## Atomic "Constants" No Longer Sacred!

If indeed the velocity of light has changed, or is changing, a certain set of other related physical "constants" has changed as well. This is a consequence that immediately follows from basic principles of physics. Setterfield and Norman did not set out to "prove" that this is indeed the case. They have, however, amassed and carefully studied a great body of data which suggests that some of most "sacred" physical constants are not constant after all. Their report is written in accord with perfectly orthodox scientific standards. That is, they have collected and analyzed all of the available data and formed an hypothesis. This hypothesis (that the velocity of light has decreased with time) is testable. It is a perfectly valid hypothesis until further data proves otherwise.

I believe it is timely and appropriate to call wider attention to this hitherto little-known investigation. The implications of Setterfield and Norman's work are profound and could make a major difference in every area of science. Mr. Setterfield has agreed to publish his scientific results in a full-length book now in preparation. Because the average reader needs a more simplified explanation than the technical report contains, and because I especially wanted to address theological issues, Barry and Trevor kindly agreed to allow me to introduce their work and explain their research findings and their implications. Norman and Setterfield's work, as I have said, presents evidence that the velocity of light has decreased with time since the universe began. This notion is not so unreasonable when we consider the history of "c."

## The Stormy History of the Velocity of Light

When the Danish astronomer Olaf Roemer (Philosophical Transactions; June 25, 1677) announced to the Paris Academie des Sciences in September 1676 that the anomalous behavior of the eclipse times of Jupiter's inner moon, Io, could be accounted for by a finite speed of light, he ran counter to the current wisdom espoused by Descartes and Cassini. It took another quarter century for scientific opinion to accept the notion that the speed of light was not infinite. Until then, the majority point of view was that the velocity of light was infinite.

The Greek philosophers generally followed Aristotle's belief that the speed of light was infinite. However, there were exceptions such as Empedocles of Acragas (c.450 B.C.) who spoke of light, "travelling or being at any given moment between the earth and its envelope, its movement being unobservable to us," (*The Works of Aristotle translated into English*, W.D. Ross, Ed.; Vol. III; Oxford Press, 1931: De Anima, p. 418b and De Sensu, pp. 446a-447b). Around 1000 A.D., the Moslem scientists Avicenna and Alhazen both believed in a finite speed for light (George Sarton, *Introduction to the History of Science* Vol. I; Baltimore, 1927; pp. 709-12). Roger Bacon (1250 A.D.) and Francis Bacon (1600 A.D.) both accepted that the speed of light was finite though very rapid. The latter wrote, "Even in sight, whereof the action is most rapid, it appears that there are required certain moments of time for its accomplishment...things which by reason of the velocity of their motion cannot be seen - as when a ball is discharged from a musket" (*Philosophical Works of Francis Bacon*; J.M. Robertson, Ed.; London, 1905; p. 363). However, in

1600 A.D., Kepler maintained the majority view that light speed was instantaneous, since space could offer no resistance to its motion (Johann Kepler; *Ad Vitellionem paralipomena astronomise pars optica traditur* Frankfurt, 1804).

It was Galileo in his *Discorsi...*, published in Leyden in 1638, who proposed that the question might be settled in true scientific fashion by an experiment over a number of miles using lanterns, telescopes, and shutters. The Accademia del Cimento of Florence reported in 1667 that such an experiment over a distance of one mile was tried, "without any observable delay" (*Essays of Natural Experiments made in the Academie del Cimento*; translated by Richard Waler, London; 1684; p. 157). However, after reporting the experimental results, Salviati, by analogy with the rapid spread of light from lightning, maintained that light velocity was fast but finite.

Descartes (who died in 1650) strongly held to a belief in the instantaneous propagation of light and, accordingly, influenced Roemer's generation of scientists, who accepted his arguments. He pointed out that we never see the sun and moon eclipsed simultaneously. However, if light took, say, one hour to travel from earth to moon, the point of co-linearity of the sun, earth, and moon system causing the eclipse would be lost and visibly so (Christiaan Huygens, *Traite de la Lumiere...*; Leyden; 1690, pp. 4-6, presented in Paris to the Academie Royale des Sciences in 1678). In 1678 Christiaan Huygens demolished Descartes' argument by pointing out, using Roemer's measurements, that light took (of the order of) seconds to get from moon to earth, maintaining both the colinearity and a finite speed.

However, only Bradley's independent confirmation published January 1, 1729 ended the opposition to a finite value for the speed of light. Roemer's work, which had split the scientific community, was at last vindicated. After 53 years of struggle, science accepted the observational fact that light travelled at a finite speed. Until recently, that finite speed has been generally been considered a fixed and immutable constant of the universe in which we live.

Scientifically speaking, the velocity of light is the highest known velocity in the physical universe. The present value has been fixed (by definition) since 1967 and is very close to 299,792.4358 kilometers per second. Almost everyone rounds this off to $3 \times 10^8$ meters/second, or 186,000 miles/second. Electronics technicians often prefer to remember the approximate number as one foot per nanosecond, the distance light travels in one-billionth of a second.

## The Velocity of Light: A Mysterious Constant

Albert Einstein made major discoveries in the early part of our century, including the well-known formula that energy, E, and mass, m, were related by the famous formula,

$$E = m\,c^2.$$

We noted earlier that the First Law of Thermodynamics says that energy and matter can neither be created nor destroyed, but only converted from one form to the other. Einstein's formulas tells us how much energy will result from the conversion of a certain amount of matter into energy (as in nuclear fission or fusion), or how much mass will result from a process where energy is converted into atomic particles, for instance. Einstein also observed that material objects grow heavier as they are accelerated towards the speed of light. In fact the mass, m, of an object travelling at velocity, v, will be greater than the rest mass $m_0$, by an amount,

$$m = m_0\,[\,1/(1-v^2/c^2)^{1/2}\,].$$

In the equation, c is, of course, the velocity of light. Einstein's relativistic increase in mass or energy of an object is not noticeable in a speeding automobile or an orbiting satellite, whose velocity is a tiny fraction of the speed of light. However, in the world-famous two-mile-long linear electron accelerator at SLAC, Stanford University, a few miles from my home, electrons injected into the accelerator reach 99.999% of the speed of light in a few feet of travel, and thereafter spend the rest of the time (and the remaining nearly two miles) getting heavier. All of the accelerator's mighty energy, inserted in the form of intense radio waves every few feet along the machine, adds total energy to the electrons by increasing their mass rather than their velocity. Thus the layman should not think that Einstein's theory is unsubstantiated by experience. Relativistic effects are common, everyday experiences to both the nuclear physicist and to the astronomer.

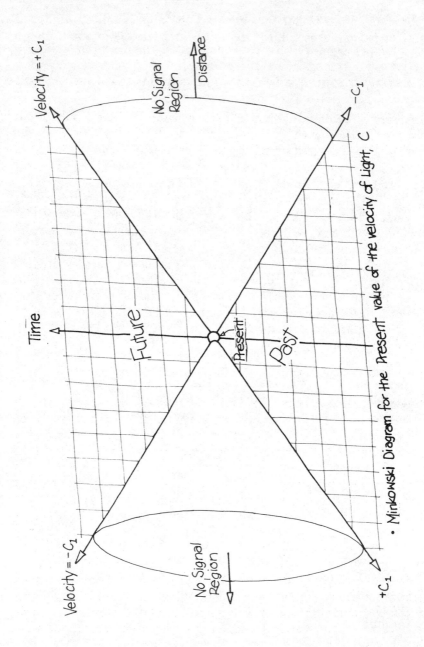

Minkowski Diagram for the present value of the velocity of Light, C

Velocity = +C₁

No Signal Region

Distance

−C₁

Time

Future

Present

Past

Velocity = −C₁

No Signal Region

+C₁

114

# Light Cones and No-Signal Regions

An interesting observation about the physical limits of our universe, arising solely because the velocity of light is finite, can be discovered with the help of what is known as a "Minkowski Diagram," reproduced on the facing page.

In this diagram, time is plotted in the vertical direction with the past at the bottom of the page and the future at the top. The intersection of the vertical axis with the horizontal axis is "the present." The horizontal axis represents distance from the observer.

The only phenomena that we are able to observe in the physical universe are events in space and time that lie outside of the two conical regions labelled "no-signal regions" since there is no way light or other information from these zones can reach us by any known physical mechanism. This "space-time" diagram is a graphic way of showing that human experience of the *physical* world is limited by the speed of light. Everything that man has experienced in the past, is now experiencing, or will perceive in the future (by means of science, for example) is limited to the regions outside the "no-signal" cones.

## Tachyons: Particles Faster than Light?

Some psychic researchers claim that psi-phenomena, (that is, "Extra Sensory Perceptions,") occur because some kind of "leakage" mechanism, presently not known to science, allows signals to cross the boundary from the "no signal" zones into the realm of the observable.

This is purely conjecture as far as I know. However, modern physics allows the mathematical invention of hypothetical particles which may or may not later prove to exist. Thus, some have suggested that particles known as "tachyons" exist in the no-signal regions. Such particles would travel faster than c, approaching c only as a lower limit. If tacyhons existed, they would (theoretically) provide a mechanism for communication faster than the speed of light. On my diagram, the lines defining the conical no-signal region, at approximately 45 degrees through the horizontal and vertical axes of the diagram, correspond to ± (plus or minus) c.

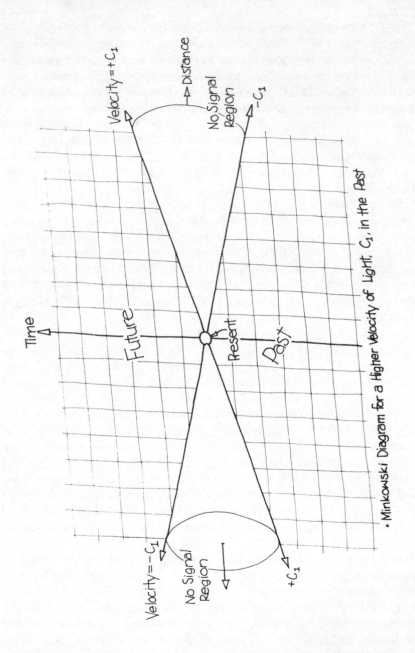

• Minkowski Diagram for a Higher Velocity of Light, $C_1$, in the Past

116

It is significant to note that if the velocity of light had been higher in the past, as I believe it was, then the size of the no-signal zones would have been *smaller* in ages past. This would allow more physically observable information into our world of experience, from what is now an inaccessible. I am speaking of physical observables not spiritual in this discussion.

## Mankinds' "Lost" Secrets

I suggest that this means that man in the past probably "saw" and "felt" and "heard" and "perceived" a richer and broader, wider and deeper range of physical reality than is now accessible to him! This idea is consistent with the Biblical view that the fall of man has been "downward" and that modern man knows less now, and is less "advanced" than his forefathers! It is also consistent with the fact that the fall of Satan and of man has "darkened" our world considerably from the state it was in at the end of creation week! The loss of ancient bodies of knowledge such as the the great library of Alexandria is thus all the more tragic for mankind. It is no wonder that many are searching for these lost secrets of the ancients in the mystery religions, the occult, and through initiation rites and ceremonies.

I have re-sketched the Minkowski diagram for a faster value of "c" on on the page opposite. Note that the size of the "no-signal region" cones has shrunk because of the higher value of c. The actual total change in the velocity of light between the time of creation and now amounts to perhaps *seven* orders of magnitude, so I have greatly understated the true difference between the two Minkowski diagrams! It is important to understand, as I have said, that in a universe where the velocity of light was higher than we know it now, more information from the created world would be accessible to our senses since the volume of the "no-signal" regions would be smaller!

In discussing the Minkowski diagram, we are talking about two regions in the *physical* world, one accessible to physical observation and the other presumably inaccessible. This is not the same thing as the division between the physical and the spiritual realms of creation. That is a different subject! As I have discussed, the fall of Satan and the fall of man have *also* affected the spiritual-physical world interface.

117

# How Two Particles Communicate

One of the perplexing problems of modern physics is how a charged particle, or a mass, communicates over distance. Since there are forces between two charged particles, or between two masses, the motion of one of these objects is "felt" by the other object. Is the signal (informing the second particle that the first has moved) sent by a virtual particle travelling at the speed of light or by a ripple in the fabric of space? The problem can be resolved if one is willing to allow the existence of a spiritual realm permeating the material universe so that information (such as prayers and answers from God) *can* travel faster than light. Although the mechanism is surely not known to us, instant action-at-a-distance probably *does* occur by means of linkages between the spiritual and material realms. God can affect two different things, miles apart, at the same instant in time, and He can order the whole by orchestrating each part all at the same time. God is an Omnipresent Spirit, in Him we live and move and have our being, so all of His sovereign power and attributes are available at every point in time and space. We have already mentioned that God often acts once from eternity causing effects to occur in past, present, and future.

## Does Modern Man Know More than his Ancestors?

Perhaps a brief philosophical and biblical digression would be in order. As a corollary to the implications of the Minkowski diagrams and the present state of the universe as one of increasing chaos, I find the Bible does teach that man has not changed since the fall, except for the intervening grace of God. Therefore, the illusion that science goes hand in hand with progress is a mistaken idea.

Every generation would like to believe we have progressed beyond the past, that evolution is occurring on a social level, if not on the physical, and that knowledge automatically leads to virtue. A "better way of life" (for ourselves) is supposed to mean we are advancing and improving on the road to perfection. These common assumptions ignore the evidence that war, crime, terrorism, and  famine are on the increase and that human mortality rates are always precisely 100%.

There is no doubt that scientific discoveries in the past century have greatly improved our quality of life, but of course the same science has also brought unprecedented possibilities for global self-destruction at the same time. As a Christian I firmly believe that history is "going somewhere," (that is, moving towards a grand consummation) and that many individuals *do* learn from the lessons of the past and, therefore, become wise in response to God's kindness and grace.

I am intrigued by a prediction made in the sixth century B.C. by the Jewish prophet Daniel who said that towards the time of the end, "**...Many shall run to and fro, and knowledge shall increase.**" (Daniel 12:4) Surely this refers to global communication, mass transportation, and the information explosion of recent years. To name just three fields: astronomy, nuclear physics and biology have all undergone enormous expansion in a few decades. Yet the wealth of new knowledge acquired has only increased the mystery found in our universe. Of course few can keep up on their own fields of speciality let alone achieve a broad point of view. But a rapid increase in knowledge these days does not mean the ancients were all ignorant and that we are superior. Biblical revelation says the very opposite!

## Nothing New Under the Sun

I do not subscribe to the notion that man as a whole is moving from primitive to civilized, or to a state of much higher collective enlightenment than that of the ancients. Primitive civilizations are the deteriorated remnants of once-great cultures, I believe, rather than cultures on their way up. I suspect, for example, that the ancient Egyptians knew many things we have not yet rediscovered. This view of things is like that held by King Solomon (c.900 B.C.) and eloquently expressed in the book of Ecclesiastes ("the Searcher") written when he was an old man. He says repeatedly that **"there is nothing new under the sun."**

**"All things are full of weariness;**
**a man cannot utter it;**
**the eye is not satisfied with seeing,**
**nor the ear filled with hearing.**
**What has been is what will be,**
**and what has been done is what will be done;**
**and there is nothing new under the sun.**

119

> Is there a thing of which it is said,
> 'See this is already new?'
> It has been already, in the ages before us.
> There is no remembrance of former things,
> nor will there be any remembrance
> of later things yet to happen
> among those who come after."

(Ecclesiastes 1:8-11)

This amazing statement suggests that human history repeats itself over and over, that the lessons of the past mostly have to be relearned every century or so. This is not to say that a Hindu view of "cyclical-time" is correct, since the Bible does present a definite progression of the ages of history in linear time. Man "progresses" towards truth only because of the intervention of God who enlightens us, one by one; otherwise all would be darkness! Progress (as we commonly think of it) is mostly an illusion and coming to realize this is, I personally believe, an important step towards true wisdom and maturity. The Apostle Paul once said, " **If any one among you thinks that he is wise in this age, let him become a fool that he may become wise."** (I Corinthians 3:18)

## The Available Measurements of "c"

To return to a discussion of the velocity of light, even though the velocity of light is very high, good measurements were made as early as 1675 A.D. by observing the eclipse times of Jupiter's satellite, Io. These times fall behind schedule as the earth draws away from Jupiter and pick up again as the earth approaches Jupiter. Since the pioneering work of Roemer using this method, the velocity of light has been measured hundreds of times by dozens of careful observers, using at least seven different methods.

All methods show that the velocity of light has decreased since 1675, though the rate of decrease has apparently slowed to nearly zero in the last three decades. When scientists make measurements, they carefully estimate every contributing source of possible error or inaccuracy and then calculate the precision of their best values and attach a ± (plus or minus) number after the main value. This method allows the observer to show that his measurement almost certainly lies midway between a high extreme and a low one. For example, Roemer's 1675 measurement reads, "307,600 ± 5400 kilometers per second," whereas a later measurement in 1875 - using the same method at Harvard University - reads "299,921 ± 13 km/sec." Of course, recent measurements are more accurate in that they have smaller error bars attached to them. For example, the 1983 laser measurement by the National Bureau of Standards in the U.S. gave "299,792.4586 ±0.0003."

Setterfield and Norman have carefully plotted almost all of the available velocity of light measurements on a curve, and they have also selected the best (most accurate) values and plotted them separately. (A very few values were not plotted in one case. These values were suspect because they have very large error bar widths compared to other measurements of the same kind in the same time period of history.) Setterfield and Norman have included the "error bars" of each measurement so as to be able to determine the limits the measurement falls between. They have also carefully re-examined the error bars themselves that were assigned by the original researchers to make certain the original measurements were realistically done in the first place, as far as it is possible for any one today to tell. What is most interesting is that both sets of c-curves (all the data, and the best data) show that the velocity of light has clearly decreased smoothly since 1675.

Next, Setterfield and Norman fed all the measurements into a computer and asked the computer to calculate a formula that best fit the data. After working a short time with the data, they discovered that the decrease in velocity of light as a function of time was well-described by a decay curve proportional to the cosecant-squared of time. A cosecant-squared curve falls steeply at first and levels off to a nearly horizontal asymptote, meaning that the velocity of light was very high early in the history of the universe but that in recent decades the subsequent reductions in c are now very small.

## Equations for Speed-of-Light Decay

Using more data and refining their calculations, Setterfield and Norman decided that an exponential and critically damped curve fit the data somewhat better. Such a curve would show an "overshoot" at late times, and indeed Setterfield and Norman believe some very recent data may show the velocity of light

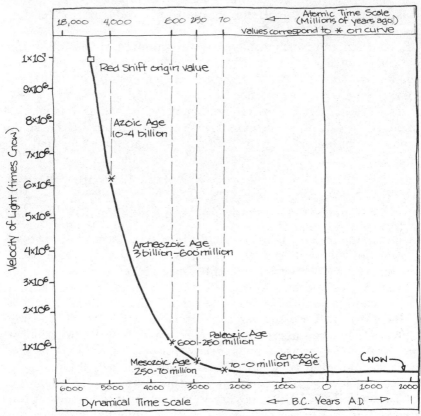

**Velocity of Light vs. Time**

Setterfield + Norman

122

increasing slightly (rebounding, as it were). Their later equation for this curve is

$$c = [\, a + e^{kt} \, (b + dt)\,]^{1/2}$$

where $k = -0.0048$, $a = 9.029 \times 10^{10}$, $b = 4.59 \times 10^{13}$, $d = -2.60 \times 10^{10}$, and t is the year A. D. A similarly excellent fit to the data is the polynomial equation

$$c = a + bT^2 + dT^8$$

Here, $a = 299792$, $b = 0.01866$, and $d = 3.8 \times 10^{-19}$. In this formula, $T = (1961 - t)$, and t is the year A.D.

Working independently, the competent and careful Canadian mathematician, Alan Montgomery, has done a statistical analysis of the known measurements on "c" and other constants of physics as well. He finds that c has been changing as a function of time with a confidence level of 95 to 99.9%. Montgomery prefers the cosecant-squared law because it can be derived analytically from physical equations whereas the two formulas above given by Setterfield and Norman are empirical fits to the observed data. Montgomery's formula is

$$c = c_{min} \text{ cosecant}^2 \, kt$$

where c is the velocity of light prior to 1958, $c_{min}$ is 299,792.458 km/sec (the year 1958 A.D.), and $k = .00025823$ radians/year; t is time in years before the asymptote, that is, before 1958. According to his data, the universe was 6083 years old in 1958.

I have redrawn Setterfield and Norman's plot of c-decay versus time (which appears as Figure 1 in the supplement to their report). My curve is shown on the facing page. It is immediately obvious from this curve that the most rapid changes in "c" occur early in earth's history. The curve also includes an approximate time scale for radioactive decay showing the comparison between dynamical and radioactive decay (atomic clock) time scales.

It is quite true that the total amount of the observed decrease in the velocity of light in three hundred years has been less than 1%, even though that amount of decrease can be shown to be statistically highly significant and highly probable. What would be most helpful would be additional "data points," evidence from the "fossil" record, or from astronomy, for example, that would give us additional measurements of c for dates earlier than 1675 A.D.

In view of the highly controversial nature of all the consequences that follow if the speed of light has truly not been a constant, it is surely important for independent investigators to check all of the existing c-data to be sure Setterfield and Norman are correct. Like any new and serious hypothesis, theirs needs careful, on-going scrutiny and it may be some years before a concensus is reached. Even then, since this work has biblical implications, I would expect heated, on-going debate by those who devote their energies to disproving the existence of God in active opposition to Him.

The above formulas show that the speed of light was only about 10-30% faster in the time of Christ than it is now, and about twice as fast when Solomon built the First Temple than it is now. In Abraham's time, (c.2000 B.C.), the velocity of light seems to have been higher by a factor of four. However, the speed of light appears to have increased very rapidly prior to 3000 B.C. It had, apparently, an initial value of about 10 million times its present value at time t = 0. This initial factor, $c_{now} = 10^{-7} c_0$, cannot be fixed with any great certainty at present. It may vary by one or more factors of ten.

Because all the known observations of the velocity of light fit all of the above formulas, it is probably scientifically more accurate to use one of these formulas showing c to be a variable than to continue to claim that the velocity of light has always been constant! Until clear, contradictory evidence comes in, it is a perfectly valid hypothesis of science to assume that the velocity of light has decayed in the past according to one of these three formulas. Accepting the data and conclusions as they are does not affect our world much for the immediate past millennium or two, as the above examples show; however, the early-time history of c-decay seems to have been very rapid. The early part of the velocity decay curve appears to be so steep we can set 8000 years as the maximum observable age of the universe.

## Many Consequences of a Changing "c"

The observed decay in the velocity of light with time is not the end of the matter, but in fact only the beginning.[5] The velocity of light is such an important "constant" of physics that it figures into

124

a large number of fundamental equations. That is, if c varies, then other important atomic constants vary also. This variation is usually proportional to c, inversely proportional to c, or $c^2$, etc. Not only have Setterfield and Norman taken the trouble to determine which atomic constants would be variable if c is found to be variable, but they have, in each case, actually gone on to show that the laboratory-measured values of those "constants" (which should vary if c varies) can be shown, from the available experimentally measured values, to also have varied with time as expected!

Thus, there is independent confirmation from measurements of other important constants as well. Combining atomic data, c-measurement data, and observed astronomical data, Setterfield and Norman show a consistent decrease in the velocity of light over the history of the universe. Furthermore, Einstein's Theory of Relativity is not affected by these conclusions. Setterfield and Norman's findings do not depend only on measurements of c: their conclusions are fully supported with a high degree of confidence (mathematically and statistically) by all other available evidence as well.

One of the most important atomic constants found to vary in proportion to c is the decay rate of radioactive nuclei! Radioactive decay data is *the* principal reason scientists have had for believing in a very old earth. It now appears that all the radio clocks have been giving times that are far older than time as measured by gravity clocks! The notion that radioactive decay rates have always been absolutely constant is one of *the most sacred cows* in physics! Yet the c-decay evidence is that the so-called geological ages of millions or hundreds of millions of years, as measured by radioactive decay processes, may be compressed into actual calendar (dynamical) times that do not exceed 7000 or 8000 years of earth history!

It is important at this point in the discussion to note that the velocity of light did not *necessarily* begin to decay at the "moment" the universe was created. In fact, I think it is likely that the universe was in existence for some length of time *before* c began to decay. It is also possible that c-decay did not *always* follow one of the above formulas: The above formulas are all based on experimental data from the past 300 years. It is, therefore, most important to search for new evidence in geology, atomic physics, astronomy, and other disciplines of science that would allow us to determine the velocity of light (or other related "constants") prior to 1675 A.D.

## The Strangeness of the Velocity of Light as a Constant

There is nothing inherent in our understanding of the physical universe that *requires* the velocity of light to be a constant. In fact, physicists, including Albert Einstein, and astute philosophers have always considered it strange and remarkable that such a constant apparently existed in the natural universe:

"We know that the discovery of the fact that the speed of light, when measured both in the direction of the rotation of the earth and in the direction opposite to that rotation, is invariable, has confronted modern astronomers with the alternative either of accepting the im-mobility of the earth or else of rejecting the usual notions of time and space. Thus it was that Einstein was led into considering space and time as two relative dimensions, variable in function of the state of movement of the observer, the only constant dimension being the speed of light. The latter would everywhere and always be the same, whereas time and space vary in relation to one another: it is as if space could shrink in favour of time, and inversely...That the movement of light is a fundamental 'measure' of the corporeal world we willingly believe, but why should this measure itself be a number, and even a definite number? ...Now, what would happen if the constant character of the speed of light ever came to be doubted - and there is every likelihood that it will be sooner or later - so that the one fixed pivot of Einstein's theory would fall down? The whole modern conception of the universe would immediately dissolve like a mirage."[5]

A number of other equally radical conclusions result from the Setterfield and Norman research. It is important to discuss some of them a bit more, hopefully in sufficiently non-technical terms so that I will not lose the average reader. As noted, a much more thorough and rigorous presentation is to be founded in the aforementioned research paper and will soon be published in book form, God willing. It may be shocking to some Christian creationists as well as to the evolutionists and the general public to declare that there is now ample scientific data to establish the youthful age of our universe and thus radically alter our view of things. Even so, I hope my readers will make every effort to satisfy their own objections and find answers to their questions by further study.

In speaking of a youthful universe, I am speaking of that portion of man's history which is accessible to our understanding through presently known laws of science and of physics. As I have said, a time discontinuity appears to have occurred at the fall of man, and before that the universe had a "pre-history" we can know in part from revelation, but not, perhaps from science.

## More than one Great Catastrophe

Many creationists have insisted on a recent creation but not given fair time and effort to reconciling the great complexity of fossil and geological records of a total time span of at most thousands of years. The c-decay data and the fossil record, I believe, require that we adopt a much more dynamic history of the earth than either side in the creation/evolution debates has supposed up till now. That is, earth has suffered *several* global catastrophes (I believe one one of them was the Flood of Noah). As a result, the fossil record shows several major species die-offs and episodes of major geological change. All of this now must be fitted into thousands of years of history, not many millions of years as previously supposed! Trends in recent years have already headed science away from the longstanding notions of "uniformitarianism" which have prevailed since the first century when the Apostle Peter confronted this same issue in his Second Epistle, Chapter 3:

"...you should remember the predictions of the holy prophets and the commandments of the Lord and Savior through your apostles. First of all you must understand this, that scoffers will come in the last days with scoffing, following their own passions and saying, 'Where is the promise of his coming? For ever since the fathers fell asleep (died), all things have continued as they were (unchanging) from the beginning of creation.' They deliberately ignore this fact, that by the word of God heavens existed long ago, and an earth formed out of water and by means of water, through which the world that then existed was deluged with water and perished. But by the same word the heavens and earth that now exist have been stored up for fire, being kept until the day of judgment and destruction of ungodly men.

"But do not ignore this one fact, beloved, that with the Lord one day is as a thousand years, and a thousand years as one day. The Lord is not slow about his promise as some count slowness, but is forbearing towards you, not wishing that any should perish, but that all should reach repentance. But the Day of the Lord will come like a thief, and then the heavens will pass away with a loud noise, and the elements will be dissolved with fire, and the earth and the works that are upon it will be burned up." (II Peter 3:2-10)

Only in the past few decades has it become acceptable and even fashionable in science to entertain notions that earth's environment has not remained uniform and favorable for life for millions of years (so as to give enough time for the slow processes of evolution to take place). Catastrophes are now admitted and sought after by scientists to explain the evidence found in the fossils and geological record of the earth.

Peter reminds us that we all readily accept prevailing views that life goes on for most of us without change and interruption. If there have been occasional disasters, these were nobody's fault and really just minor flaws in nature. God is not so interested in man, as some claim, or else He would surely have prevented them! The notion that God actually intervened previously in history to destroy an entire civilization is something we would surely prefer not to think about, especially when Peter's warning is directed at us who live at the close of a second age when another even more terrible judgment is about to fall on a corrupted mankind. Only those who take refuge in Christ, the Greater Ark of safety, can escape.

"Since all these things are thus to be dissolved, what sort of persons ought you to be in lives of holiness and godliness, waiting for and hastening the coming day of God, because of which the heavens will be kindled and the elements will melt with fire! But according to his promise we wait for new heavens and a new earth in which righteousness dwells." ( II Peter 3:11-13)

# Notes to Chapter Six

1. See Barrow, John D. and Tipler, Frank J. *The Anthropic Cosmological Principle* (Oxford University Press; New York, 1986). This book thoroughly discusses the fact that important constants of physics, relevant to the atomic scale of the very small, (the microcosm) also appear in equations governing the behavior of the galaxies and largest scales known in the universe (the macrocosm). It seems to be no accident to the authors that man lies in the center of the universe and that he falls midway between the very small and the very large which encompass about 40 orders of magnitude, total. A wonderful book on the scale of relative sizes found in the universe is *Powers of Ten* by Philip and Phylis Morrison and the office of Charles and Ray Eames (Scientific American Library; W.H. Freeman and Company; San Francisco; 1982).

2. *The Atomic Constants, Light and Time* by Barry Setterfield and Trevor Norman; Dept. of Mathematical Sciences; Flinders University, South Australia. Invited Research Paper prepared for SRI International; Menlo Park, California. (Available air mail post paid for $10 total from the authors; Box 318; Blackwood, S.A. 5051; Australia).

3. Alan Montgomery; 29 Radford Dr.; Ajax, Ontario, Canada L1T 2G5 (personal communication).

4. A number of objections and critical questions were raised by Setterfield's work in response to an earlier report now out of print: *The Velocity of Light and the Age of the Universe*, published in 1983 by the Creation Science Association of South Australia. In Volumes 1 and 2 of the Australian Creation Society Journal, *Ex Nihilo*, Setterfield has carefully answered most of his critics to my satisfaction. The problems with accurately correcting carbon dating are complex and are the subject of an article *Carbon-14 Dating, Tree-Ring Dating, And Speed of Light Decay (A Preliminary Model)*, by Barry Setterfield, **Ex Nihilo Technical Journal**, Vol. 2., 1986. It is simply impossible for me to repeat these arguments for the reader of this book and still cover the theological issues remaining to be discussed. Some of these issues are, unfortunately, temporarily out of print. It is hoped they

will either be reprinted and made available once again, or that Mr. Setterfield's new book (now in preparation) will incorporate these answers to critics and additional comments. *Ex Nihilo* is available by subscription from Master Books; P.O. Box 1606; El Cajon, CA., 92022. For a historical summary of other scientific papers, since 1924, suggesting that some of the fundamental constants of physics have varied, see William R. Corliss, *Mysterious Universe: A Handbook of Astronomical Anomalies*, (Sourcebook Project; P.O. Box 107; Glen Arm, MD, 21057).

5. Titus Burckhardt, *Mirror of the Intellect: Essays on Traditional Science and Sacred Art* (State University of New York Press; Albany, 1987), pp. 27-28,.

**Chapter Seven**

## Changing Constants in Physics ?

## Some Further Technical Matters

In the last chapter, I discussed the evidence that points to the likelihood of a radical decrease in the velocity of light since creation. It is quite true that the rate of decrease appears to now be nearly zero, so that careful ongoing measurements of "c" may not tell us much. However, I think it could prove possible to re-examine the "fossil" records of the past, or the light from the most-distant stars for indirect evidence of the actual velocity of light at a given calendar (dynamical) time in the past. This is because other atomic "constants" are variable if it is true that c is a variable. This follows as a direct consequence of the laws of physics.

How do we determine which of the other "constants" is truly constant and has invariant with the passage of time? In their analysis, Setterfield and Norman have insisted that the Conservation Laws (The First and Second Laws of Thermodynamics) are immutable. Almost every scientist feels a great reluctance to abandon these basic "laws" of physics since they have proven so reliable and useful in recorded history.

## The Stretching-out and Relaxation of Space Itself

An even more important question is "Why has the velocity of light decreased with time?" One equation of interest, (which is part of James Clerk Maxwell's famous equations for electromagnetic wave propagation) says that the velocity of light is related to two properties of empty space known as the "magnetic permeability," $\mu_0$, and the "electrical permittivity," $\varepsilon_0$, according to the equation,

$$1/c^2 = \varepsilon_0 \, \mu_0.$$

Setterfield and Norman have shown that the permittivity of free space has not changed with time according to the best available measurements. Therefore, it is evidently the *permeability* of free space that has varied inversely with $c^2$. This parameter, the magnetic permeability, is apparently related to the stretching out of free space at the time of creation. When God stretched out the "firmament of the heavens" as described in Genesis, the value of $\mu_0$ had its lowest value.

Sometime after creation, the heavens were apparently "released" from their initial stretched-out condition, much as one would let air out of a filled balloon. A "relaxation" effect has set in, since creation, perhaps beginning when Satan or man fell. This relaxation has allowed the universe to begin to shrink (not to continue to expand as the Big Bang theory has erroneously claimed).

The shrinkage of free space has been the cause, apparently, for the observed slowing down of the velocity of light . At first, the collapsing of the universe was very rapid, but for the past 2000 years or so, the rate has been slow and nearly zero since about 1958 - if the data and analysis are correct.

Also, from James Clerk Maxwell's electromagnetic theory, we can calculate the so-called impedance of free space (commonly used in antenna design). The present value is 377 ohms, and the exact formula is

$$Z = [\, \mu_0/\, \varepsilon_0 ]^{\,1/2}.$$

The impedance of free space tells us how radio waves, or photons of light, will travel through space. Z also gives us the ratio of the electric field vector, **E**, to the magnetic field vector, **H**, in free space. The refractive index of any medium - whether empty space or other material (the property of a glass lens, for example, which enables it to bend a beam of light) - is given by the formula,

$$n = \varepsilon_0 \mu_0 \, [E/\, H]$$

where E is the magnitude (strength) of the electric field vector and H that of the magnetic field vector. It follows that optical path lengths everywhere in the universe have been changing since the relaxation of the universe set in, some time after creation. This result has a number of consequences for astronomy and the actual diameter of the known universe.

The energy of a photon can be calculated from Einstein's equation, $E = m\,c^2$. If we use this formula, it is easy to see that the photon has "apparent mass", m. Photon energy is also known to be equal to hf, where h is Planck's constant and f is the frequency of the emitted light of the photon. The energy of a photon can also be expressed in terms of wavelength, $\lambda$, rather than frequency,

$$E = h c / \lambda = h f.$$

Thus Planck's "constant" should be found to be varying with time, inversely proportional to c. This is borne out by a table of data given in the Setterfield and Norman report. Their table indeed shows an increase in Planck's constant with time with a high degree of statistical confidence. Setterfield and Norman have shown that the wavelength of radiation, at the time a radio-wave or light photon is emitted, is invariant for constant energy. However, once a radio-wave leaves the source, or a photon departs from its parent atom, energy and momentum are apparently both conserved. Also the product "hc" is a true constant which does not vary with time. Photons are mysterious "particles"and it has always been assumed that they travel at the "constant" speed of light. They are even assumed to not exist at lesser speeds! How photons behave in flight, if c is not constant, is at present a bit uncertain in my own mind, at the time of this writing.

In their report, Setterfield and Norman show that the deBroglie wavelengths for moving particles and the Compton wavelength are c-independent. The energy of an orbiting electron, the fine-structure constant, and the Rydberg constant are also shown to be c-independent and thus truly constant with time. The gyromagnetic ratio, $\gamma = e/2mc$, is found to vary proportionally to c.

As I have said, the wavelength of light emitted from atoms, (for instance, the atoms on a distant star), can be shown the be independent of any changes in c. However, the relative energy of the *emitted* light wave is inversely proportional to c, and if c decreases while the light wave is on its journey, its energy and its momentum must be conserved in flight. The intensity of the light, related to the wave amplitude, increases proportionally to c. Thus there should be proportionally less dimming of light from distant stars. In order for light to maintain energy conservation in flight, as c decays, the frequency of the emitted light must decrease inversely proportionally to c. The relaxation of free space, causing the observed c-decay, and increasing optical path-length, occurs everywhere in the universe at the same time. This results in a corresponding shift in the light wave's energy towards longer wavelengths, creating the so-called "red-shift".

133

# The Universe is Not Expanding?

According to Setterfield and Norman, the universe can also be shown to be *contracting* - and not expanding as previously supposed - that is, the red-shift is due to the fact that light from distant stars has decreased in velocity since it was emitted. If the universe were collapsing and the speed of light constant, astronomers would observe a "blue shift" of light from distant stars, rather than a red one. The Setterfield data seems to show c to has been decreasing *more rapidly* than the actual physical collapse of the universe, giving light a net red-shift when it reaches our telescopes. The velocity of light was very likely a constant for a certain period of time (or epoch) before c-decay commenced, I believe, on theological grounds. This would affect the present observed background radio and optical temperatures of space as Setterfield and Norman discuss in their paper. Their study of the latter data suggests to them that the initial value of c was perhaps some ten million times its present value. This value of $10^7$ is presently uncertain by one or more factors of ten (orders of magnitude). This is not a serious matter; it only shows how little we know at present about how and why c has decayed.

The energy, E, associated with a mass, m, is $E = m\ c^2$, as stated earlier. This means that the mass of an object varies as $1/c^2$. Measurements show that the ratio e/mc has been exhibiting a (least-squares fit) decrease of 679.0 EMU/gm with a 99.17% confidence level, quoting from Setterfield and Norman's report. However, when gravitational effects are brought in, it can be shown that *rest-masses are invariant with respect to c-decay in their own time frames*, whether dynamical (gravitational) or atomic. When rest masses are measured dynamically as in the ratio of e/mc, they show the above-noted variation. Conversely, when dynamical phenomena are measured atomically, we note a variation in the Gravitational Constant, G. Just to summarize a few other conclusions reached in the Setterfield and Norman study: the Bohr magneton, gas constant $R_0$, and Avagadro's number, $N_0$, all can be shown to be c-independent.

# Newton's Law and the Gravity Metric of Space

As far as gravity is concerned, the gravitational force, F, between objects of mass $m_1$ and $m_2$ is given by Newton's formula,

$$F = G\, m_1\, m_2 / r^2,$$

where G is the Gravitational constant and r is the distance between the objects. Space has built-in gravitational properties similar to its electrical properties mentioned above. This gives rise to the so-called "Schwartschild metric for free space", which also is related to the stretched-outness of free space, and finally affects one of the most mysterious constants found in astronomy, the famous Einstein Cosmological Constant, $\Lambda$. Setterfield and Norman's work appears to show that the cosmological constant, $\Lambda$, is negative, rather than positive as previously assumed..Thus, the so-called "missing mass" needed in the Big Bang theory to cause an expanding universe to contract is not now of concern, because the universe has apparently been contracting already since early times.[1] The fact that the rate of decrease in the velocity of light has been nearly zero in the past few decades indicates that the contraction of the universe has presently dropped to nearly zero, though of course red-shifted light from distant stars is still reaching our telescopes. None of us need fear, therefore, that the universe is about to become a black hole.

A look at the way gravitational equations are affected by c-decay shows that the motion of the planets about the sun and all gravity clock rates remain *unaffected by c-decay*. Thus, the "correct" way of telling time in our universe is to rely on gravity clocks, not atomic clocks. This is consistent with Genesis Chapter One.

## Radioactive Decay Rates Have Changed

The decay of c with time has been shown by Setterfield and Norman to affect, for example, the speed nucleons in the atom move in their orbitals, and the alpha particle escape frequency, $\lambda^*$, is shown to be proportional to c. This means, in short, that all *radioactive decay rates have decreased in direct proportion to c* throughout the history of the universe. The general conclusion is simply that the radio dating methods, whether carbon-14, potassium-argon, or any other atomic-clock method which appear to give us very great ages, must all be corrected to take into account the change in decay rates of the atomic nucleus.

Up until now, radioactive decay rates have been believed to be "sacred" constants. This assumption - upon which so much modern cosmology and modern nuclear physics rests - can now be questioned and a new model built, taking into account the observed variations in c and related "constants" of physics. The curve of c-decay vs. time, given in the last chapter, includes a radioactive-decay time scale to show how very significant the change in atomic clock-time has been since creation week. As stated earlier, the main reason scientists have had, till now, for believing the universe is very old has been the widespread acceptance that all radioactive decay processes were invariant (unchanging). If the velocity has really decayed according to the Setterfield and Norman study, the true age of the universe as we know it scientifically shrinks to less than 8000 years. I have already noted that there are dozens of other reasons for believing in a recent creation, all of which are usually ignored by modern science without having been refuted of challenged seriously!

The details of the research I have sketched briefly in this chapter have already been studied thoroughly for the past seven years by my distinguished colleagues and reported by them. My short summary above is not intended to overwhelm the layman, but to call attention to what I believe is the great scientific importance of the Setterfield and Norman model so that the scientifically-inclined will order and read their report carefully and examine it with a fine-toothed comb.

I have set out to write a theological book, not to rewrite modern physics or cosmology. I have purposed to show that a biblical view of the universe need not be compartmentalized and divorced from reasonable science. As a Christian, I find Setterfield and Norman's ideas are worth considering seriously. I am not attempting to force their data or conclusions to fit my own preconceived notions of creation. I think the tasks of looking afresh at all the implications of c-decay must now be undertaken by others. Of course, I will be temporarily disappointed if fatal flaws are found in the velocity of light data, or the Setterfield and Norman conclusions that c has in fact decayed; however in the meantime the subject deserves diligent study. So far I have seen no serious reason presented to stop pursuit of these new and refreshing ideas.

In the long run science cannot be at odds with the Bible, nor with the very character of God The course of history has shown that eventually there is more harmony, not less, between science and the Word of God. God's revelation of Himself in nature will eventually be clear enough to convince everyone!

In Chapter 2, I pointed out that major shifts in scientific theories often come from careful examination of very small effects previously overlooked by others. The work of Setterfield and Norman *may* well be an excellent example of that very process at work in science. Perhaps it is sufficient to point to all the available data and propose the hypothesis that the Setterfield-Norman data are correct and build from there. If the resultant theory fits other data, all the better. Like all really competent scientists, Setterfield and Norman would be perfectly willing to be proven wrong. If they are correct, and I personally have thrown my bets in alongside theirs, then many new mysteries in science remain to be investigated.

❂

# Notes to Chapter Seven

1. P. C. W. Davis, a prolific author, has written an excellent, easy-to-understand book *The Accidental Universe* (Cambridge University Press; New York, 1982). The subject is modern cosmology; the practical meaning of the Einstein cosmological principles are made about as clear as they can be for the layman. He also discusses "weak" and "strong" anthropic principles, which (as I said earlier) concern the fact that the atomic constants appear also in the equations of astronomy, effectively coupling the sub-atomic realm with that of the galactic scale of things. With man midway between the world of the very small and the very large, then somehow man's presence in the universe may not be an accident. A slight alteration in any number of small terms would mean a universe in which life as we know it could not live, or would not be here at all. I believe that man's behavior as God's steward over nature *does* affect nature on all scales.

Man is not an "accident" of nature or of evolution, but a creature of spirit and of dust, embedded into the fabric of the universe such that his actions affect everything else around him. On the other hand, man is not free to destroy himself and the rest of creation because God in His sovereignty is in full control of all that takes place in heaven and on earth. Much of God's activity in nature takes place apart from man. God allows evil to run its full course; however, His final intervention in the affairs of men will come in time, of that we are assured.

# Chapter Eight

## Flawed Men (and Women)

## Eastern and Western Views of Life

I have mentioned that science cannot and does not operate in a vacuum. Surrounding science on all sides are philosophical and religious presuppositions. Science as a body of acquired information also has a history associated with it, and we cannot ignore the individuality, uniqueness, and belief systems of the thousands of scientists who have helped to build the edifice. Science as we know it today is the product of Western civilization and contains very many inputs from Biblical thought whether Jewish or Christian. This does not change the fact that most modern scientists are non-religious or have discarded the belief systems of their predecessors while keeping their derived knowledge.

I believe man to be incurably religious by nature. Pure materialism is never very successful, but sooner or later is replaced by some kind of religious outlook on life. In fact, the Latin roots of the word "religion" literally mean "to bind back again"; that is to find once more the roots from which we have sprung, or to discover our unique place in the cosmos. Even if we do not believe in an external God who is more intelligence than ourselves, we eventually come to the belief that we ourselves ought to take on the role of god in the absence of a better alternative!

In recent years some scientists[1] (a minority, I believe) have looked to Eastern religion or Eastern philosophy for further clues to push back the frontiers of physics. I am not opposed to Eastern philosophy in preference to Western, for I believe both are, at heart, complementary ways of looking at things, much as male and female persons perceive truth and set their priorities in different ways, both of which are valid!

I believe all of mankind has been periodically enlightened by God's grace and much truth remains in other religions of the world. If sorted out properly, such truth can enrich the impoverished, dried-up soul of Western, scientific, super-rational man. Eastern mysticism, however, does differ in many ways from Biblical thought, and it is therefore not useful to science to attempt to "solve" certain dilemmas of modern physics by searching through mystical or pantheistic notions of reality.

It might seem at first that concepts borrowed from Chinese philosophy would not be especially parallel to ideas found in the Scripture. However, there is evidence in the Chinese language that their earliest civilization knew about Biblical (Old Testament) revelation. Most world religious and philosophical traditions contain *some* truth and reveal earlier periods of greater enlightenment. Thus, it is sometimes helpful to examine Biblical truth from the starting point of kernels and concepts found outside the Christian tradition. Eastern philosophy in many ways complements a Western point of view and can be profitable to study since the Bible itself is more of an Oriental book than many in the West have supposed.

## Yin, Yang and the Tao

In Chinese philosophy, the interesting concept of yin and yang applies to more than the relationship between the sexes. For example, in Taoism, heaven is masculine and earth, feminine, suggesting the dependence of the entire creation upon the Creator. A whole series of possible interactions between the yin and yang in life is contained in the Chinese Book of Changes, the *I Ching*. This is not a book Christians should "follow," although it contains wise sayings resembling the Proverbs of Solomon.

In using the *I Ching* the proper set of "wise sayings" selected from the text (when confronted by a particular situation) is determined by tossing a coin or throwing yarrow sticks. Thus, the underlying principle is chance, or a belief in oracles. Carl Jung elaborated on why the *I Ching* "works" and seems to give helpful answers, by supposing that apparently unrelated events are governed by "synchronicity" or by the "collective unconscious" of mankind. However, these concepts are easily extended to a form of Eastern pantheism which in actuality denies the God of the Bible as the One who is in charge.

God does not prohibit our quest for knowledge *except* in matters of the occult where we come too easily into bondage to "the elemental principles (or spirits) of the world" or fall under the influence of spiritual power exercised by the "prince of the power of the air," the "god of this present age." Our Lord has, therefore, specifically prohibited our consulting fortune-tellers, Tarot cards, astrologers, or familiar spirits. I personally believe that consulting the *I Ching* is in this category. Nevertheless, the book is fascinating to read because it hints at harmony as well as disorder in life and shows the complexity of choices and the paradoxes we often face in making choices. As James says, there are two types of wisdom in the world; mere knowledge is never enough to guide us along the way: **"Who is wise and understanding among you? By his good life let him show his works in the meekness of wisdom. But if you have bitter jealousy and selfish ambition in your hearts, do not boast and be false to the truth. This wisdom is not such as comes down from above, but is earthly, unspiritual, devilish. For where jealousy and selfish ambition exist there will be disorder and every vile practice. But the wisdom from above is first pure, then peaceable, gentle, open to reason, full of mercy and good fruits, without uncertainty and insincerity. And the harvest of righteousness is sown in peace by those who make peace."** (James 3:13-18)

## The Tao and the Sabbath Rest

The word "Tao" (pronounced "dow") is not easily translated but roughly can be thought of as "the way" or "the union of opposites by the middle way" (Jung).

"The flowing stream is a key metaphor for Taoism. One is to 'flow with' the Tao, or life-principle of the universe. To be 'with it,' and not 'against the Tao,' is to be fulfilled. Not rigidity and aggressiveness, but flexibility and simple naturalness are encouraged. To 'get ahead' it is best to lie low and keep quiet. The curse of life is the many layers of artificiality of which cities are the symbol: the impersonal relationships; the dictated tastes of custom; the abstractions of bureaucracy and technological civilization. Wisdom resides with the simple, uncomplicated people who are close to the soil and who move with the rhythms of nature; not the wordy professors, the politically ambitious 'climbers,' and the scheming, overweight moneylenders. To live in harmony with the Tao - the really real - is the path to authenticity."[2]

The Tao implies the principle that man can choose to live in harmony with God, with nature, and with himself so as to find peace of mind and inner wholeness, or he can elect the path of disharmony and destruction.[3] The Tao is impersonal in Chinese philosophy since Taoism does not claim to be a religion; however the Christian equivalent to "the path of harmony in life" is the Biblical idea of the Sabbath *rest*: The writer of the letter to Hebrews says, **"whoever enters God's rest also ceases from his own labors** (strivings) **as God did from His."** (Hebrews 4:10) In the New Testament "the Way" is a Person, not a concept: Jesus said, **"...*I am the Way,* and the Truth, and the Life; no one comes to the Father, but by me."** (John 14:6) If then an individual is willing to come into partnership with Jesus, wholeness and peace with God (*shalom*) are the end result: **"Come to me, all who labor** (strive by self-effort) **and are heavy laden, and I will give you rest. Take my yoke upon you, and learn from me; for I am gentle and lowly in heart, and you will find rest for your souls. For my yoke is easy, and my burden is light."** (Matthew 11:28-30)

The well known symbol (above) of the yin and yang, two fish encircling one another, is a picture of the complementary nature of the sexes, their need for one another and their interdependence. Harmony between the sexes is not only a problem for the married: half of life is best understood from the feminine point of view and the other half by the masculine - both poles are needed to get at the whole "great mystery" of life. "Complementary" is a better word to use than "opposites" in describing the differences between the sexes - and the Bible does teach equality of the sexes. The generic term "man" or "brethren" in Scripture, unless otherwise stated, means "men and women" because the two sexes are the same in spirit, the innermost core of our being. **"But he who is united to the Lord becomes one spirit with Him."** (I Corinthians 6:17) speaks of the believer's union with Christ whether that person is male or female. The yin/yang nature of things is apparently related to the creation of the first man, Adam, who was formed from the dust and became a living soul when God breathed the breath of life into his nostrils. At this point yin and yang were evidently in some sort of perfect union - Eve had not yet been taken out of Adam.

## Both Man and Woman Bear the Image of God

What exactly is meant by Adam/Eve jointly carrying the image of God is not a simple subject to discuss.[4] The fact that there are wide differences of opinion and of Biblical interpretation on this subject should surprise no one in today's world! Also, God, Elohim, is a *triune* being, not *dual* (M/F) in nature, as man is, so that the compound unity of the relationships of the Persons in the Godhead is more complex than the concept of yin and yang can convey even in a rudimentary way.

Although Adam named the animals (who were evidently already separately male and female), **"there was found no helper suitable for him."**

**"So the LORD God caused a deep sleep to fall upon the man, and while he slept took one of his ribs and closed up its place with flesh; and the rib which the LORD God had taken from the man he made into a woman and brought her to the man. Then the man said, "This at last is bone of my bones and flesh of my flesh; she shall be called Woman, because she was taken out of Man.""**
(Genesis 2:21-23)

Since the separation of the man Adam into two individuals took place before their fall, it is legitimate to suppose that perfect harmony originally prevailed between the man and the woman, and between each of them and God. Surely they were functioning, for a season, in harmony with the creation God had given them to subdue and govern as stewards.

## Disharmony Between the Sexes

No one knows how long this state of innocence prevailed, and it is probably meaningless to ask such a question since *historical* time, as we know it now, appears to be measured from the fall. No children were born to Adam and his wife until after the fall, so most Jewish commentators believe that the temptation, fall, and expulsion from the garden followed soon after "creation week."

The fall of man was so total and so profound that nature itself was affected: death and decay began to spread through the universe and even certain laws of physics seem to have been affected. One of the most terrible consequences of the fall, for us, was surely the catastrophic dysfunction between the man and his wife, between the yin and the yang. Apart from God's intervening grace, this disruption would have been permanently fatal for the entire human race.

The fall did not only throw into disarray the relationships between the man and the woman, their God, and all their offspring. It also affected Nature, causing disharmony, disorder, and decay, that is, Nature was "subjected to futility." Thousands of years of human experience now vividly testifies to the complexity of the interrelationships between the sexes and even the great difficulty of achieving such things as a happy marriage without hard work and God's grace:

"Marriage is not comfortable and harmonious; rather, it is a place of individuation where a person rubs up against himself and against his partner, bumps against himself in love and in rejection, and in this fashion comes to know himself, the world, good and evil, the heights and the depths."[5]

God's dealings with the man and his wife after their joint-fall reveal that there were terrible consequences for the both of them and for the created order because they ate of the fruit of the tree of the knowledge of good and evil. However, God immediately gave them gracious new provisions and promises leading one day to a new creation, a new race, a new family of mankind. That new family would attain to higher, grander, and greater levels than the old creation could ever know. This possibility arises not only because of the provision of a Savior, but because fallen men and women would thereafter freely choose to love God, or freely choose to reject Him. Thus arises the awesomeness of free will which would make man more like His creator than ever before.

Various Bible commentators have noted that the finest and most marvelous aspects of God's Person cannot be manifested in an unfallen world. These include grace, mercy, forgiveness, justice, healing, and self-giving love. God did not cause the fall, but He allowed it; and even before creation, He planned to save and restore His creatures through His own suffering.

Strictly speaking, it is not proper to think of life as half-male and half-female or to suppose that these two halves make one whole. Discussing the obvious differences between the sexes may be the best way to ease into the subject of male and female genders and roles:

"Sexuality, which penetrates our whole being, will not be expressed physically in the resurrection body, but it will have its expression at the soulish and the spiritual levels. God has a purpose for it in the life to come. That is why we are given physical sex. It is designed to teach us what we are like, who we are, what our role is, in the life to come. Male organs are external to emphasize, as one of the marvelous visual aids that God is always employing, that the male role is one of visible leadership. He is designed to take the initiative, and yet to do so with tenderness and gentleness...Female organs are internal, hidden, to indicate the role of women as being inwardly sensitive, far deeper emotionally than men, more subjective, contributing deeper insights than man ever does, having a greater sense of compassion, and responding to that which leads. All this is designed to teach us truth about our relationships with one another and with God himself. That is why, throughout the Scriptures, God appears in relationship to the Christian as the lover, the aggressor, the male. We are the bride, the responders, the followers, and that is consistent through the Scriptures."[6]

## Masculinity and Femininity Compared

In discussing human behavior, it is easier to speak of masculinity and femininity rather than maleness and femaleness because every man and every woman has both "masculine" and "feminine" traits. In the male, masculine priorities are dominant and on the surface (or ought to be) and feminine "qualities" are secondary and "recessive." In women the opposite is true. Men who suppress their feminine heritage (which they receive from their mothers, as it were), men who live out only masculine qualities are cold, insensitive, calculating, domineering, overly rational, harshly aggressive, and too self-assertive. Totally "feminine" women, on the other hand, are thought of as wall-flowers in their passivity, governed by emotions or intuition and hopelessly dependent, unable to take the initiative in the smallest matters. It is difficult to discuss such things without constructing a model, so for the sake of discussion, I have made my own list of those qualities that I think men are

145

naturally good at and those that women by nature find easily perform:

## Masculinity:
Paternal Instincts
Leads
Initiates
Values logical thinking
Logos-centered mind
Bases decisions on reasoning
Assertive, aggressive, and outgoing towards the world
Provides outside support and protects (as a soldier)
Feels ill-at-ease in passive, receptive roles
May undervalue importance of emotions and intuition
Specialist in"Doingness","Withoutness" (hunts, thrusts, ranges)
Mechanistic, scientific, digital-thinking
Divided left and right brain

## Femininity:
Maternal Instincts
Responder
Follower
Values emotions highly
Heightened intuitive ability and perception
Sophia-based thinking
Shelters, nurtures, heals
At ease in a passive, receptive role
Takes leadership with difficulty when male leader defaults
Natural concern for the present quality of life
Specialist in "Beingness" and "Withinness"
Nestles, nurtures, shelters, traps
Artistic, creative, analog thinking
More unified left and right brain

It is immediately clear that a yin/yang relationship between many of these complementary sets of priorities exists in every one of us. It is therefore important for each and every individual to become whole in relationship to God as requisite to healthy relationships in life such as marriage. To know ourselves, however, means to successfully know and relate to the opposite sex, to live in harmony with the other's values and priorities. This is true whether one is single or married, divorced or widowed. For example, when a man gets in touch with his (largely unconscious) femininity, he ought to become more loving, gentle, receptive, and emotional. When a woman discovers her inner masculine heritage and accepts these qualities, she should develop poise, dignity, self-confidence, and stature. In either case, the man is said to be more manly and the woman, more regal in stature, attaining character traits everyone recognizes and admires as marks of a well-integrated personality.

Unfortunately, stereotyped views of what men are supposed to be like and especially the "proper" role and place of woman as the "inferior sex" abound in our society. These views are rightly objected to by thinking men and women acquainted which the many examples of manhood and womanhood God endorses for us in the Bible. So much teaching, preaching, and leading in the churches is all done by men that the feminine voice of truth is suppressed or undervalued. This great imbalance in church life virtually precludes healing, freedom of spirit, and the expression of the beauty God has built into His church.

## The On-Going Pattern of the Fall

To cite but one example from the Bible of how yin and yang work together either constructively or destructively within a given individual, James says this about temptation:

**"Let no one say when he is tempted, 'I am tempted by God,' for God himself tempts no one, and he has no experience in evil things. But each person is tempted when he is lured and enticed by his own desire. Then desire when it has conceived gives birth to sin; and sin when it is full grown, brings forth death."** (James 1:13-15 My paraphrase)

Here is the pattern of the fall itself repeated in our lives every day. Eve was deceived, but Adam was not deceived. The desires and yearnings of the fallen femininity in each of us are appealed to by the tempter who plays upon the "masculine," logical, reasoning mind to get us to rationalize and justify what we know to be wrong so that we hand over our wills, temporarily, to the tempter's lie and to our inner desires. In so doing, the yin and yang within us fall into that disharmony which constitutes spiritual adultery, and a "child" of death is conceived in us! (A literal illustration of this verse is the death of the child born to Bathsheba after King

147

David's adultery.) When a man or woman gives in to the "fallen femininity" of the world (that is, to the appeals of "the great harlot") the result is sin and death.

Conversely, standing firm, relying on the indwelling power of God to resist and overcome temptation, reverses the fall, allowing God to redeem both "fallen masculinity" and "fallen femininity" and to restore, as it were, the lost yin/yang harmony or the Tao. To understand this better, we need only think of the situation of Adam in the garden after Eve had fallen but he had not. Adam's choice to sin was deliberate, but we can imagine a situation where Adam (as a type of Christ) withstood temptation and became the instrument for redemption of Eve.

Surely the most helpful and wonderful book on masculinity and femininity in the whole Bible is the Song of Solomon. This great book of Hebrew poetry can be read and understood on at least three levels. First, it is a love story of a man and woman who grow and mature as their relationship progresses. Secondly, Jewish and Christian commentators have always acknowledged that the Song pictured Yahweh's love for Israel or Christ's love for His church. Watchman Nee in his commentary[7] takes the view that the woman in the Song of Solomon typifies every individual believer, as well as the church collectively, and therefore gives us a picture of discipleship seen from an "interior" point of view. Yet a third interpretation is possible: Solomon is a male Hebrew name derived from **shalom,** but Shulamite is a feminine name, *also* derived from **shalom.** "Shalom" in Hebrew means that peace which comes from being a whole person, to be at peace with God and with himself. Thus, we can consider that Solomon and Shulamite are the same person seen from two different points of view (one outward and one inward), and the book is therefore a study of personal wholeness in relationships with God.

The latter view is one that C. G. Jung's psychology was at home with. He described the "coniunctio oppositorum," or "union of opposites," as the goal of individuation. Jung noted that the "mysterium coniunctio," the relationship between yin and yang, can be "true" or "false."[8] Jung did not hold to a Biblical view of evil, but he certainly struggled with evil as a reality in human affairs. In his own way, Jung attempted to understand how we all must come to terms with the masculine and feminine principles in life, and in God Himself, in order to be whole persons, whether we are single or whether we are married.

# The Last Adam and His Bride

Jesus, the last Adam, could have been expected to marry and raise a family. It was unusual in Israel for a single man not to be married by the age of 30. From Scripture it is clear that Jesus suffered all possible forms of temptation known to man, "yet (He was) without sin." He did not marry because His primary purpose in coming was to atone for the sin of mankind on the cross, not to raise a family in the usual sense.

After returning from the dead and ascending to heaven, Jesus put into effect a New Covenant for living. Under this New Covenant (having greater and better promises than those contained in the Old Covenant), many Old Testament rules and regulations were set aside. Certain customs and traditions such as circumcision and keeping the Sabbath were discarded. But no change was made in the oldest human institution, marriage. In fact, the status of marriage is elevated in the New Testament because it is to be compared, as "a great mystery," with the relationship between Christ and His church.

At this point, Christian theology goes even further beyond the limitations of Chinese philosophy as it has been handed down to us today. Jesus came to build a new race, and He came to take a Bride, the Second Eve. The marriage is not to be a physical one, but a spiritual one; and the Bride, not merely one woman. The bride is the church: the company of those redeemed in this present age - men, women, children numbering many millions! The relationship of Christ and His church *transcends* concepts as yin, yang, and the Tao.

The figure of the Bride of Christ is but one of Seven Figures of the church found in the New Testament. The church is also called the "Body of Christ," with Christ as the Head. The latter figure of the church is more or less a "masculine" picture of the church - the church in its relationship outwardly to the world. When Jesus spoke of building His church He said, "...**I will build my church, and the powers of death shall not prevail** (be able to withstand) **against it.**" (Matthew 16:18) The church is an invincible, conquering army in its outward activity in the world.

When the church gathers around the Lord, the object of her worship and devotion, she is the Bride of Christ, feminine in her submission, obedience, and responsiveness. Thus, the church is outwardly masculine but inwardly feminine and contains

149

balanced attributes of both men and women, since both sexes constitute her membership rolls.

The church, militant and triumphant, is properly described by a verse in the Song of Solomon describing the Shulamite in her maturity: **"Who is this that looks forth like the dawn, fair as the moon, bright as the sun, terrible as an army with banners."** (Song of Solomon 6:10) The sun is an ancient symbol of Christ, the greater light who rules the day, and the **"sun of righteousness who comes with healing in its wings."** The moon is a symbol of the church which is in itself barren and lifeless but rules the night by reflecting the light of the sun, **"until the day breaks..."**

## Is There Sexuality in Heaven?

Some have inferred that the distinctions between the sexes would disappear in heaven since there we **"...neither marry nor are given in marriage..."** (Luke 20:35); however, the opposite is surely true: the differences will be enhanced and made more splendid. C. S. Lewis, among contemporary writers, makes this point clearly in his Science Fiction Trilogy and other works. One reason the delights and pleasures of heaven are at present denied us is that we are unable to avoid the excesses of legitimate joys and pleasures in our fallen condition and in this present world. Rather than enjoying a little wine, we are all too eager to drink to excess and lose self-control. Forgetting that God can only bless sexual expression in marriage, we ruin our lives with destructive, promiscuous behavior, preferring the **"... pleasures of sin for a season."** (Hebrews 11:25) (KJV) Not content with God's promises to provide for our material needs generously, we insist on striving after wealth and power forgetting that **"whoever wishes to be a friend of the world** (system) **makes himself an enemy of God."** (James 4:4)

In heaven the restraints can be removed safely to allow us ecstatic pleasures without fear of sin. In such a state the differences between the sexes will be enhanced and completed:

"...Some kind of procession was approaching us, and the light came from the persons who composed it. First came bright Spirits, not the Spirits of men, who danced and scattered flowers--- soundlessly falling, lightly drifting flowers, though by the standards of the ghost-world (life as we know it now on earth) each petal would have weighed a hundred-weight and their fall would have been like the crashing of boulders. Then, on the left and right, at each side of the forest avenue, came youthful shapes, boys upon one hand, and girls upon the other. If I could remember their singing and write down their notes, no man who read that score would ever grow sick or old. Between them went musicians; and after these a lady in whose honour all this was being done. I cannot remember whether she was naked or clothed. If she were naked, then it must have been the almost visible penumbra of her courtesy and joy which produces in my memory the illusion of a great and shining train that followed her across the happy grass. If she were clothed, then the illusion of nakedness is doubtless due to the clarity with which her inmost spirit shone through the clothes. For clothes in that country are not a disguise: the spiritual body lives along each thread and turns them into living organs. A robe or a crown is there as much one with the wearer's features as a lip or an eye..."9

Introducing this fanciful picture of what heaven may be like, Lewis writes,

"You cannot take all luggage with you on all journeys; on one journey even your right hand and your right eye may be among the things you have to leave behind. We are not living in a world where all roads are radii of a circle and where all, if followed long enough, will therefore draw gradually nearer and finally meet at the centre: rather in a world where every road, after a few miles, forks into two, and each of those into two again, and at each fork you must make a decision. Even on a biological level life is not like a pool but like a tree. It does not move towards unity but away from it and the creatures grow further apart as they increase in perfection. Good, as it ripens, becomes more different not only from evil but from other good. I do not think that all who choose wrong roads perish; but their rescue consists in being put back on the right road. A wrong sum can be put right: but only by going back till you find your error and working it afresh from that point, never by simply going on. Evil can be undone, but it cannot 'develop' into good. Time does not heal it. The spell must be unwound, bit by bit, 'with backwards mutters of dissevering power'---or else not. It is still 'either-or.' If we insist on keeping Hell (or even earth) we shall not see Heaven: if we accept Heaven we shall not be able to retain even the smallest and most intimate souvenirs of Hell. I believe, to be sure, that any man who reaches Heaven will find that what he has abandoned (even in plucking out his right eye) was precisely nothing: that the kernel of what he was really seeking even in his most depraved wishes will be there, beyond expectation, waiting for him in 'the High Countries.'"[10]

I have written about the flawed relationship that now exists between the sexes, and about one of the Eastern paths, Taoism, to show that East and West are now largely in disharmony just as man's fall has disrupted nature and marred the whole of creation. Only occasionally do we glimpse things as they might have been, or ought to have been. In insisting on its "superior" point view, "Western science" leaves out much truth about man and his universe to construct its models!

## Totally Depraved?

It is not pleasant news for us, at first, to learn about the human condition from the Bible, for man is "fallen," and in the eyes of a Holy God, the fall was into a state of total depravity. This is not to say that man, the bearer of God's image and likeness, is without worth, for he has infinite worth as the object of God's mercy and love. Yet the prophet Jeremiah said about the human condition, **"The heart is deceitful above all things, and desperately corrupt; who can understand it?"** (Jeremiah 17:9)

Perhaps it seems strange to the reader to introduce masculinity, femininity, and the human moral, ethical condition into a discussion which has mostly to do with the science and the nature of the created universe. However, because God is a Personal God, and because the universe was made for man (according to the Bible), how man is faring in his relationship with his Creator has a lot to do with the state-of-the-union message about nature - and the heavens and the earth as a coupled system.

All evil does not reside with man. In fact, evil began with a revolt of the angels, not with an independent rebellion by man. Evil is present right now in heaven as it is on earth, and God's plan for reversing and undoing the effects of the fall has not yet been fully worked out in history for all to see.

It should be obvious to all that a good deal of what is wrong in the world has to do with ourselves. Part of the problem is our own blindness to our moral condition and to our need for deliverance and healing.

We surely therefore need revelation more than science to understand ourselves. A good place to start is the Epistle to the Romans in the New Testament which opens with this description of mankind: (To further clarify this passage, I have footnoted the text and included definitions of Greek words obtained from several good Greek lexicons and Bible Encyclopedias.)

"...the wrath[11] of God is revealed[12] from heaven[13] against all ungodliness[14] and wickedness[15] of men who by their wickedness suppress[16] the truth. For what can be known about God is plain to them, because God has shown it to them. Ever since the creation of the world his invisible nature, namely, his eternal power and deity, has been clearly perceived in the things that have been made. So they are without excuse; for although they knew God they did not honor him as God or give thanks to him, but they became futile in their thinking[17] and their senseless minds[18] were darkened. Claiming to be wise, they became fools, and exchanged the glory of the incorruptible God for images resembling mortal man or birds or animals or reptiles.

"Therefore God gave them up in the lusts[19] of their hearts to impurity[20], to the dishonoring of their bodies among themselves, because they exchanged the truth about God for a lie[21] and worshiped and served the creature rather than the Creator, who is blessed for ever! Amen.

"For this reason God gave them up to dishonorable passions[22]. Their women exchanged natural relations for unnatural, and the men likewise gave up natural relations with women and were consumed[23] with passion[24] for one another, men committing shameful[25] acts with men and receiving in their own persons[26] the due penalty for their error.

"And since they did not see fit to acknowledge God, God gave them up to a base mind[27] and to improper conduct. They were filled with all (kinds of) unrighteousness, fornication[28], wickedness[29], covetousness[30], malice[31]. Full of envy[32], murder[33], strife[34], deceit[35], malignity[36], they are gossips[37], backbiters[38], haters of God[39], insolent[40], haughty[41], boastful[42], inventors of evil things[43], disobedient to parents[44], foolish[45], faithless[46], heartless[47], implacable[48], unmerciful[49]. Though they know God's decree that those who do such things deserve to die, they not only do them but approve those who practice them. (Romans 1:18-32 My translation)

153

"...all men...are under the power of sin, as it is written:
'None is righteous, no not one;
no one understands, no one seeks for God.
All have turned aside, together they have gone wrong;
no one does good, not even one.'
'Their throat is an open grave,
they use their tongues to deceive.'
'The venom of asps is under their lips,'
'Their mouth is full of curses and bitterness.'
'Their feet are swift to shed blood,
in their paths are ruin and misery,
and the way of peace they do not know.'
'There is no fear of God before their eyes.'"
(Romans 3:9-18)

## Curses and Blessings

Having looked briefly at the condition of mankind since the exipulsion from Eden, I would like to suggest a few more of the interrelationships that exist between man and creation as described in the Bible. For example, Hebrew thought includes both curses and blessings, man to man or God to man. A long list of both curses and blessings is found, for example, in Deuteronomy 27 and 28. A curse is said to carry with it the power of enforcement so that it remains effective until lifted.

In reading the Old Testament, it is clear that famines (brought about by the power of God to deny rainfall) or other conditions of nature that are harmful to human prosperity are God's instruments of punishment and chastisement upon mankind. It is not necessary for us to find the direct causal mechanism by which God uses natural forces to judge mankind. The very forces of judgment that come via nature and even their timing could just as easily be built into the laws of nature by an All-powerful, All-wise, All-knowing God. This does not reduce God to an impersonal, detached, uncaring remote position of non-involvement in human affairs. The God of the Bible is far more than a First Cause who sets everything into motion and then stands back like a bystander to see how things work. Pantheism takes an opposite - and also false - position by stating that God is the sum total of all that is.

154

I mentioned in Chapter Four that we can assume from the Bible that a group of the angels fell before man fell, so that there was active and destructive evil at work in creation prior to the fall of man. Thus, the creation *could* have suffered from marring and scarring due entirely to the activity of Satan and his host that occurred *before* the fall of man.

But there seems to have been no "curse" *on the earth* until the fall of man. The fall of man came when Lucifer approached Eve in her innocence with his deceptive plan to undo the old creation, prevent the new, and and render useless and ruined, men who are made in the likeness of God. The New Testament says that Eve was deceived but Adam was not deceived and became the transgressor, so that the moral accountability for the fall is Adam's. This is confirmed in several New Testament passages (such as Romans 5:18, I Corinthians 15: 22, 45).

Eve, having fallen, but still attractive and beautiful to Adam, became the seductress who led her husband into sin. She is therefore a type of the figurative woman in Scripture known as the "great harlot." Adam - knowing full well the consequences of his deliberate disobedience - yielded to his wife. In fact, he was apparently not paying attention in the first place, or he could have stopped his wife from eating the forbidden fruit and prevented both her fall and his.

As described in the Bible, death is a complex process, having entered the world through man: "**...sin came into the world through one man (Adam) and death through sin, and so death spread to all men because all men sinned.**" (Romans 5:12) Spiritual death means that man's spirit has become disconnected from that real Life that only God possesses. (The absence of the spirit from the body altogether means physical death according to the Epistle of James.) In addition to the physical aging and dying of the body, spiritual death brings boredom, meaninglessness, depression, anxiety, despair, and other symptoms as we continue living apart from God. If we do not seek reconciliation with God in this life, spiritual death is followed by physical death, and then by judgment and banishment forever from the presence of God. This last condition is known in Scripture as "the second death."

Those who receive Jesus into their lives as Lord immediately gain eternal life although the body remains unredeemed until the resurrection. Jesus said, "**...he who believeth in me, though he were dead, yet shall he live: and whosoever liveth and believeth in me shall never die.**" (John 11:25,26) (KJV) Scripture frequently treats those who do not yet know Christ as already dead; for example Paul says we are "...**dead in trespasses and sins.**" (Ephesians 2:1) until we come to Christ.

## Consequences of the Fall

After their fall, God proceeded immediately to restore Adam and Eve to fellowship with Himself, but first came a series of judgments, or "curses." First, God addressed the serpent, then the woman, and last, the man. Although the human sin was forgiven, cause-and-effect consequences were inevitable. Permanent changes in the created order also resulted that would further handicap man is his future assignments and endeavors. Not only did Adam and Eve become subject to death they also became constrained to live out the remainder of their days in a narrow "time corridor" before "staging" into resurrection bodies. Adam and Eve had been designated lords of nature, so their fall was serious indeed, not only for themselves, but because they had lost the wisdom and power to serve as good stewards over creation.

Thus, both the animate and inanimate became "cursed"; that is, subject to decay and purposelessness. Before the fall, when God walked with Adam and Eve in the garden, the power of God flowed freely from the spiritual world into the physical world. Everything was "very good," and the universe was probably not running down in the sense of using up energy resources and becoming more disordered. That is, I believe that the Second Law of Thermodynamics *may* not yet have been in effect. It is also *possible* that there was, at first, no radioactive decay of any atomic nuclei, for reasons to be explained later.

Genesis, Chapter 3 describes the fall and its consequences in highly poetic but multi-level imagery. Numerous commentators on Genesis have elaborated on God's statement to the man:

"**...Because you have listened to the voice of your wife, and have eaten of the tree of which I commanded you, 'You shall not eat of it,' cursed is the ground because of you; in toil you shall eat of it all the days of your life; thorns and thistles it shall bring forth to you; and you shall eat the plants of the field. In the sweat of your face you shall eat bread till you return to the ground, for out of it you were taken; you are dust, and to dust you shall return.**" (Genesis 3:17-19)

156

An interesting commentary on this curse over creation is given by the Apostle Paul in the 8th Chapter of his Epistle to the Romans in the New Testament. Paul writes,

**"I consider that the sufferings of this present time are not worth comparing with the glory that is to be revealed to us. For the creation waits with eager longing for the revealing of the sons of God; for the creation was subjected to futility, not of its own will but by the will of him who subjected it in hope; because the creation itself will be set free from its bondage to decay[50] and obtain the glorious liberty of the children of God. We know that the whole creation has been groaning in travail** (labor-pains) **together until now; and not only the creation, but we ourselves, who have the first fruits of the Spirit, groan inwardly as we wait for adoption as sons, the redemption of our bodies."** (Romans 8:18-23)

This passage says that the entire universe has been affected by the fall of man. Sand has been thrown into the delicate gears of the once-flawless, perfect mechanism of creation. Not only does this cause sickness, disease, war, rust, decay, wearing-out and growing old but also energy depletion and a drift towards chaos.

The whole world system is also becoming less and less manageable by men and by government. It is running ever more out of control towards a disaster that only intervention by God can resolve. In addition, Satan has assumed rulership over territory that does not belong to him, namely the world-system. His plan is to completely wreck and ruin mankind by any means at his disposal. The subject of "the Angelic Conflict," discussed briefly in Chapter Nine, impinges on human affairs in the physical world more than is commonly realized in our materialistically-minded mind society.

Far from building "the great society," fallen man is the great polluter and he carelessly disrupts delicately-balanced natural forces which he poorly understands. Lest he cause even more damage, man's territorial bounds and powers have been limited by His Creator until the coming time of liberation and restoration, (Acts 17:24-31). However, it should be obvious that human activity has been affecting the weather, depleting the ozone layer, raising the sea levels, destroying natural life in the oceans and on land, poisoning the atmosphere, and harmfully disturbing both himself and his environment in countless ways. Even "beneficial" inventions of man often have unexpected and harmful side-effects.

## Evil in the Heavens and on the Earth

In attempting to understand our universe better, we should allow for the discovery that there are probably two kinds of "damage" that have been occurring to our once flawlessly-perfect and beautiful universe. The first type of havoc has been and is being wrought on man and nature alike by fallen angelic beings under the leadership of one Jesus referred to as "...**a murderer from the beginning and** (who) **has nothing to do with the truth...**" (John 8:44) It may well be that evidence of this kind of destructive activity might be found in certain events in the heavens observed by astronomers. On earth the direct activity of Satan can be seen wherever open idolatry, occult practices, and magic arts are deliberately encouraged. Otherwise, his activity is more concealed, deceitful, and indirect.

The second source of on-going destruction comes from man himself. When anyone of us acts from self-centeredness, he is in fact empowered by the Evil One, "...**the spirit that is now at work in the sons of disobedience.**" (Ephesians 2:2) The world-system (that is, the social order) is also under the dominion of Evil at the present time. James, Chapter 4, connects all interpersonal conflicts, marriage failures, gang fights, and even wars among nations to the hedonistic life styles of men. James says that a cause-and-effect relationship exists between the pursuit of a self-centered, pleasure-seeking human life style and the inevitable consequence of disharmony.

However, creation is not running out of control. The New Testament repeatedly states that the well-qualified man, Jesus, is running all affairs - cosmic and local. Jesus sets the limits and bounds of our habitations and our powers. He constantly intervenes, and He is now directing history towards the goals that were set before the worlds were ever made. Christ's final direct and personal intervention will spare those who belong to Him from final destruction.

## God and Matter

Several passages in the New Testament especially seem to apply to the structure of matter and how it is held together by God in a dynamic way. One passage, already discussed, is the prologue of the Book of Hebrews:

"In many separate revelations - each of which set forth a portion of the Truth - and in different ways God spoke of old to [our] forefathers in *and* by the prophets, [But] in the last of these days He has spoken to us in [the person of a] Son, Whom He appointed Heir *and* lawful Owner of all things, also by *and* through Whom He created the worlds *and* the reaches of space and the ages of time - (that is) He made, produced, built, operated and arranged them in order. He is the sole expression of the glory of God - the Light-being, the out-raying or radiance of the divine - and He is the perfect imprint *and* very image of [God's] nature, upholding *and* maintaining *and* guiding *and* propelling the universe by His mighty word of power..." (Hebrews 1:1-3) (Amplified Bible)

This passage indicates that Jesus, the Son of God, now has an active role in sustaining the old created order - since He "guides, propels and sustains" the universe by His "mighty word of power." Here, we are shown that God is active in and through all the forces of nature at the present time.

As I have said, I believe that in general God does not seem to be interfering too much with the running-down and the wearing-out of the old creation, as discussed earlier. By "not interfering" I mean that God ordinarily works through the existing laws of nature and physics and that we do not seem to be able to observe these laws being broken by natural or supernatural causes.

Rather, God is evidently concentrating His energies on the building a new humanity and preparing a new creation ("new heavens and a new earth") which will raise the redeemed of earth to a higher and greater state than Adam and Eve enjoyed in Eden. The relevant passage in Romans, Chapter 8 says that the "curse on creation" resulting from the fall, will be lifted when Jesus returns and not before.

Writing to the Colossians, the Apostle Paul says this of Jesus:

"He is the image of the invisible God, the first-born of all creation; for *in him* all things were created, in heaven and on earth, visible and invisible, whether thrones or dominions or principalities or authorities - all things were created through him and for him. He is before all things (beyond, outside of), and in him all things hold together. He is the head of the body, the church; he is the beginning; the first-born from the dead, that in everything he might have be pre-eminent. For in him all the fullness was pleased to dwell, and through him to reconcile to himself all things, whether on earth or in heaven, making peace by the blood of his cross. And you, who once were estranged and

159

hostile in mind, doing evil deeds, he has now reconciled in his body of flesh by his death, in order to present you holy and blameless and irreproachable before him, provided that you continue in the faith, stable and steadfast, not shifting from the hope of the gospel which you heard, which has been preached to every creature under heaven, and of which I Paul, became a minister." (Colossians 1:15-23)

## Jesus Holds the Universe Together

A key word in this passage is surely the Greek word *sunistemi* which means "to stand-together," "to be compacted together," "to be constituted with." I believe this passage can be applied to the structure of the atom, for example. The nucleus of every atom is held together by what physicists call "weak" and "strong" forces. (Physicists are familiar with four basic forces in the natural world: gravity, electrical forces, a "strong," and a "weak" nuclear force which act at very short ranges. The first two forces decrease in strength inversely with the square of the distance between two objects. Very recently two additional close-range, weak gravitational forces have been suggested. These are thought to be quantum mechanical corrections to Newton's Law of Gravitation.)

The nucleus of the atom contains positively-charged and neutral particles, to use a simplistic model. Mutual electrostatic repulsion between the like-positive protons would drive the nucleus apart if it were not for the "strong force" which binds the nucleus together. I believe that the strong force may have decreased in intensity (that is, weakened at the time of the fall) causing some atomic nuclei to become unstable and to start radioactively decaying. If this hypothesis is correct, radioactive clocks may not have started running at the "moment" of creation or during the "seven days of creation" but later on. This hypothesis further complicates a comparison of radioactive decay dates with gravity clocks during the early days of the universe.

## God and Nuclear Forces

The third New Testament passage which talks about atomic structure and physics is II Peter:

160

"But the day of the Lord will come like a thief, and then the heavens will pass away with a loud noise and the elements (atoms) will be dissolved with fire and the earth and the works that are upon it will be burned up." (II Peter 3:10)

The Greek word translated "elements" in this passage is *stoicheia* which means the building blocks of the universe, or "the ordered arrangement of things." It can also mean the "atomic elements." The word translated "dissolved" is literally (in Greek) the word "unloosed." This suggests a further, future letting-go of the nuclear binding force that holds the nucleus together. This passage strongly suggests (to my way of thinking) that the active power of God is behind the mysterious strong force that holds every atomic nucleus together. Of course, if this is so, all the other fundamental forces of nature are likewise forces that originate with Christ and His sustaining direction of the old creation.

If this is a correct view, were God to merely relax His grasp on the universe every atom would come apart "by fire" (that is, by nuclear fire). God dynamically sustains the universe, including the atoms themselves. They are "stable" only because force from the spiritual realm is being supplied into the physical nuclear binding fields. Whatever we may think of God and physics, the Bible leaves us with no room to doubt that God does care about the sparrow that falls to the ground, the widow, the orphan, and the homeless. He does not lose track of His children and watches over them with infinite, patient, intimate Fatherly care. He sustains the universe by His mighty word of power. He also alters the status quo and, in response to prayer, frequently changes the course of entire nations.

Man is flawed but so also is creation. The laws of physics operative today are the laws of a fallen world, and many of these "laws" appear to have been discontinuous at one or more points of time in the past. It is also possible that the laws of nature suffered further, downhill degradation at the time of the Flood of Noah, as discussed in Chapter 10.

As we look at our present world and think of the Biblical Apocalypse which lies ahead, we cannot help but be aware that man brings God's judgments upon himself through his sinful actions in rebellion against God. Some of these "judgments of God" are built into the system, made a part of the very laws of physics, so they seem to us to be "natural." In fact, such judgments are "Acts of God."

# Accidents and Disasters in the Plan of God

Almost everyone of us, especially when we are young, thinks himself especially privileged, a favorite child of God, who is exempted from trouble while others are not. The daily news causes us distress to be sure, and eventually we all must cope with automobile or plane crashes, family tragedies, and the possibilities of nuclear winter and global genocide. In the prevailing existential vacuum of our day many live solely for today's pleasures and give no thought to misfortune, loss of health, accident, or death. The notion that there are no absolutes and that moral standards are relative ignores the Biblical evidence that a holy, just, loving God is to be found behind the scenes of all that happens in our world. We do not live in a meaningless universe driven by blind forces, by mere probability without an underlying guiding Hand. God has not lost control of any details nor is the devil winning, as some suppose.

A recent issue of National Geographic contains vivid pictures of the sudden destruction of Pompeii, a reminder that we easily forget the disasters that have already overtaken whole cities in earth's history. Many scoff at the flood of Noah, though Jesus Himself said that it had occurred. And each of us faces death for which we must always be prepared in an uncertain and evil world. A terrible plane crash, a great earthquake disturb us all because they remind us that our life spans are limited and that (like it or not) we might, any of us, "die before our time" according to the plan and providence of God.

The problem which secular writers and most contemporary scientists are unwilling to address is the problem of evil: how and why do human beings behave so self-destructively (all of us) and what is the solution.

The biblical answer is that human life is in the grip of an alien foe, a destroyer, and that his death grip is broken only when Jesus sets up His inner kingdom in a man's heart. Because God is holy, which implies not only purity but also wholeness, only a suitable blood sacrifice makes possible the building of a bridge from God to fallen man. That Jesus was an important historical figure and teacher few deny. That He is alive and able to heal can be discovered only by calling on Him for help. This simple and humbling step has never been known to leave the seeker

162

unrewarded - that is the simple testimony of millions of believers over thousands of years of history.

## The Words of Jesus about Disasters

Jesus did comment on disasters that befall us for no apparent reason. He gave a startling answer to the question "Why did God allow such a terrible thing to happen?" Luke 13 gives the account. Jesus was asked about a tragedy in the temple in Jerusalem when Pilate had brutally killed some devout Jewish worshipers at the time of the temple daily sacrifice, mingling their blood with that of the sacrificed animals: **"...Do you think that these Galileans were worse sinners than all other Galileans, because they suffered thus? I tell you, No; but unless you repent you will all likewise perish. Or those eighteen upon whom the tower** (under construction) **in Siloam fell, and killed them all, do you think they were worse offenders than all the others who dwelt in Jerusalem? I tell you, No; but unless you repent you will all likewise perish."** (Luke 13:1-5)

The word "repent" is an unpleasant word to our modern ears - no doubt partly due to some of the now-discredited radio and TV church Bible thumpers we have all grown up with. Not all of these preachers have been men of the Word, or men of good repute. The Hebrew words for repentance mean simply to change one's life style. If one is a liar, he is to stop lying; if guilty of stealing, he is to make restitution; if in need of forgiveness, he is to seek forgiveness. And, he is to be accountable to others for his actions.

The New Testament Greek word translated "repentance" means "to have a different mind"; that is, to see things in a different light. Of course, when we are "enlightened," our behavior changes for the better. Repentance means radical change - within and without - in anyone who comes to know God.

Disasters can happen to individuals or entire societies because of our general and individual alienation from God. Not a single one of us deserves to be rescued; not one of us is worthy because of meritorious performance. Were God just, without being merciful, all of us would be lost. That He rescues any at all is the amazing thing. In fact, Planet Earth is headed for near self-destruction as many different writers in the Bible remind us, but it is always in our hands to change the warp and woof of the tapestry of history by our free will choices. The coming dark time of trouble on earth spoken of by the Hebrew prophets (including Jesus the Chief Prophet of both Israel and the Church) will be, at the same time, the dawning of a new golden age for mankind. But the words of Jesus are binding on all men: "...**unless you repent, you will all likewise perish.**" (Luke 13:5) Because God loves us all, there is hope. The future of mankind is to be nothing short of paradise regained. But it is possible to be left out.

## Notes to Chapter Eight

1. Fritjof Capra, *The Tao of Physics* (Shambhala Press; Boulder, 1975); Zukav, *The Dancing Wu Li Masters* (Bantam New Age Books; New York, 1979); David Bohm, *Quantum Theory* (Prentice-Hall, Englewood Press; N..J., 1951). See also his *Wholeness and the Implicate Order*. Also recommended is F. David Peat, *Synchronicity: The Bridge Between Matter and Mind* (Bantam New Age Books; New York, 1987), and Fred Alan Wolf, *Star Wave: Mind, Consciousness and Quantum Physics* ((Macmillan Publishing Co.; New York, 1984).

2. Paul Clasper, *Eastern Paths and the Christian Way* (Orbis Books; Maryknoll, NY, 1982) and C. George Fry, James R. King, Eugene R. Swanger and Herbert C. Wolf, *Great Asian Religions* (Baker Book House; Grand Rapids, 1984).

3. Alan Watts, a former Episcopal priest who became a skillful teacher of Eastern religions and philosophy until his death in 1973, wrote lucidly for Western minds on these subjects. His book, *Tao: The Watercourse Way* (Pantheon Books; NY, 1975), is a good introduction to Taoism and Zen Buddhism.

4. Paul Jewett gives a thorough discussion of various possible models in his book, *Man as male and female* (Eerdmans; Grand Rapids, 1975).

5. Adolph Guggenbühl-Craig, *Marriage, Dead or Alive* (Spring Publications; Zurich 1977).

6. Ray C. Stedman, *Expository Studies in I Corinthians* (Word Books; Waco, Texas 1981).

7. Watchman Nee, *Song of Songs* (Christian Literature Crusade; Fort Washington PA., 1965).

8. Jolande Jacobi, *The Psychology of C. G. Jung* (Yale University Press; New Haven, 1973).

9. *The Great Divorce* (Macmillan Press; NY 1946) pp. 106-107.

10. Preface to *The Great Divorce*

11. The wrath (*orge*) of God is His abiding and constant anger directed against human evil. Another Greek word (*tthumos*) is used to describe outbursts of anger commonly known to us all, which flare up and subside. Both words are used in Revelation 16:19 and 19:15 to describe the "fierceness of the wrath" of Almighty God.
12. The wrath of God is continuously being poured forth and "rests" (or "abides") upon those who do evil. This divine resistance towards human evil is actively at work quietly and silently so as to produce the deterioration in quality of life described below.
13. Here heaven refers to the invisible dimension of the spiritual world which surrounds us and in which we live and move. God is not remote, detached, and uninvolved in our lives - regardless of whether or not we know Him and serve Him.
14. ungodliness: *asebeia*, means "without *eusebia*," godliness. The word means without reverence to God as well as active in opposition to God's purposes.
15. wickedness: *adikia*, means "not conforming to that which is right"; (*a+dike*), means, without righteousness. A comprehensive term for wrong doing.
16. Suppress, hold down, or repress the truth, or hinder it. All men know there is a God, and they also know He is just and righteous. From childhood this knowledge is resisted and buried by each one of us. The social institutions of society work with us in this conspiracy to deny God in all aspects of life.
17. Empty or futile speculations or reasonings. These include the philosophies of society, much of what is called science, and the governing principles we set for our own lives when we ignore

God's standards. Actually, our ignoring God puts into bondage to evil while God desires for us, "...the glorious liberty of the sons of God." (Romans 8:21)

18. Literally, "their hearts were darkened without understanding." The heart is the inner man, the center of what and who we are. Jesus said, "For out of the heart come evil thoughts..." (Matthew 15:19)

19. Or, God handed them over to the power of sin. Lust is *epithumia*, strong desire; in this case, a reaching out for pleasure, especially passionate desire for forbidden things (Barclay). Rejection of God means we are open to error yet blinded to what is actually happening to us. This portion of the letter to the Romans refers to "mankind" - not specifically to individual men and women, although one person's personal rejection of God always sets him or her on a path of moral decay with one or more of the consequences described inevitably occurring. Moral decay of society can occur gradually over several generations, resulting finally in outbreaks of immorality that are no longer hidden or private and breakdowns of family life and family values. Increased terrorism, crime, and lawlessness follow. Note that the process that leads to total moral decay begins when men cease to worship and serve God for who He is. All men are without excuse since knowledge of God is available to all men. The first step in the downward spiral is to place another person first in one's affections instead of God. After that, our progressively-increasing idolatry causes us to be enslaved to the love of money, power, prestige, selfish sexual gratification, etc. We lose our dignity and humanity in the process.

20. uncleanness: *akatharsia* means literally "not having been purged or cleansed." Our English word catharsis comes from the Greek root.

21. Literally, the lie, not "a" lie. The lie is the widely-held notion men have that they can handle life on their own without God. It is the belief that man is master of his own fate and destiny and is capable of being god of his own life. This was the lie presented to Eve in the garden.

22. Behavior which dishonors the body, "passions of dishonor." The word passions here is not *epithumia* but *pathe*, which is the more passive word meaning the diseased condition out of which the lusts spring (Barclay).

23. consumed: *exekauthesan*, literally means "burned out." Obsessive cravings that cannot be satiated, or desires that can

166

never be fulfilled and only increase to produce a wasted, devastated, ruined condition.

24. passion, or lust: **orexis** means "reaching out" (beyond proper bounds of moral restraint or decency) in order to appropriate something for oneself. This particular Greek word is used here only in the New Testament, .

25. unseemly: that is, not suitable, not proper, or unbecoming. Related to disfigurement (Vincent) and contrasted with gracefulness.

26. The penalty which is due from God as a consequence of violating the Divine Law. Older words for homosexuality are "perversion" and "inversion" which imply a twisting or reversing of sexual identity within oneself. That part of the personality which should be turned inward towards God becomes twisted outward producing a confused sense of masculine or feminine sexual identity.

27. base mind: **adokimon noun** literally means a mind that cannot stand the test. Vincent says (noting the play on words in the Greek), "As they did not approve, God gave them up unto a mind disapproved."

28. Greek, **porneia,** from which we get our common word pornography. The word refers to all forms of sexual activity outside of marriage, especially sex between unmarried persons. Illicit sex is associated with moral uncleanness and impurity of the heart and the affections.

29. Greek, **poneria,** from the Greek root word meaning "to toil." The word has come to mean active, outgoing, malignant wickedness, as opposed to kind, gracious, serviceable - hence, destructive, injurious evil (Vine). The desire of doing harm to others, to assault, to corrupt the innocence or goodness of others. To drag others down to one's own level of destructive evil (Barclay).

30. covetousness: **pleonexia** is the desire to have more, always in a bad sense; i.e., wanting more than one needs or more than one can use, craving something another possesses for selfish purposes, etc.

31. malice or evil: **kakos** means defective or evil in character, base; as opposed to fair, advisable, good in character, beneficial, useful (Vine). Grasping for money or goods, regardless of honor or honesty; ambition which tramples on others to gain something for oneself; unbridled lust which takes its pleasure where it has no right to take, (Barclay).

167

32. envy: *phthonos* means a feeling of displeasure produced by hearing of the gain or prosperity of others, always in the evil sense.

33. murder: *phonos* means to take the life of another out of hatred.

34. strife: *eridos* means contentiousness, an argumentative attitude. The contention which is born of envy, ambition; the desire for prestige, place, office, and prominence - coming from a jealous heart, (Barclay).

35. deceit: *dolos* means crafty, full of guile, ensnaring. The Greek word is used to mean to debase precious metals or adulterate fine wine. The quality of deceit in a man who can no longer operate in a straightforward way but resorts to devious, underhanded ways to get his own way (Barclay).

36. malignity: *kakoetheia* means malicious disposition, malevolence. An evil disposition that tends to put the worst construction on everything (Vine). To suppose the worst about other people, to place wrong interpretations on innocent actions.

37. whisperers: *psithuristes* means those who spread secret slander.

38. backbiters: *katalalos* means to speak against the character of another person, especially when he or she is not present.

39. haters of God: *theostuges* means hateful to God.

40. insolent: *hubristes* means violent, despiteful. The attitude of a man when he is so proud he defies God, fate, and fortune. Extreme self-confidence. An attitude which takes devilish delight in inflicting mental and physical anguish on others (Barclay).

41. haughty: *huperephanos* means showing oneself to be above others or over them; i.e., disdainful, arrogant, proud, the opposite of humble and lowly. "A certain contempt for everyone except oneself" (Barclay).

42. boastful: *alazon* means "one who wanders about the country," a vagabond, hence an imposter (Vine). The stock word for quacks who sell fake medicines, health foods, remedies, etc. Pretending to be someone who one is not, a braggart.

43. inventors of evil: *epheuretas kakon* means to invent, contrive. or seek out new ways of committing evil.

44. disobedient to parents: unpersuadable, obstinate; rejection of the will of God with regard to how one should treat one's parents.

45. foolish: *asunetos* means without understanding in a moral and spiritual sense. Lacking in spiritual perception or dis-

cernment. The man who will not use the mind and brain God has given him (Barclay).

46. faithless: *asynthetous* means covenant-breaking; that is, refusing to honor agreements, pay debts, or keep promises.

47. heartless: *astorgous* means without natural affection, especially within one's family.

48. implacable: *aspondous* means "without a libation"; hence, without a truce. One who cannot be persuaded to enter into an agreement. Unreasonable, treacherous.

49. unmerciful: *aneleemon* means without mercy. Without pity, placing a low value on human life or another person's distress or rights.

50. decay: *phthora* means destruction occurring by means of, or because of corruption.

**Chapter Nine**

# Jesus, Lord of the Angels

## The Inhabitants of the Spiritual Realm

Science has paid slight heed to anything that is really "there" in the spiritual realm, to information that is gained by revelation, and this is quite proper in one sense. But science and research have come to be thought of as *the* way of solving everything, and that is a most serious error to believe and live by. The physical world is only "half" of the creation, and the invisible half is the more permanent and substantial whereas the physical world is the world of shadows.

Ultimately, all energy and power and life come into the physical world from the unseen spiritual world even though we do not know how to formulate scientific laws to describe the interactions between these two realms. Since science is bounded by many intrinsic limitations, the responsibility for our failure to be fully-informed and educated concerning the dual realms of creation should probably be laid at the door of the church for failing in its task of instructing us in areas that are, by nature, out-of-bounds for science.

I began this book by emphasizing that, according to the Bible, there are *two* levels of creation: one physical and one (the more important) spiritual. The two realms were coupled before the fall and remain coupled now, but imperfectly so. Both realms have become flawed, subsequent to creation, and active evil is present in both realms. Scientific observations of the physical world have been, are now, and will be affected by happenings in the spiritual world. Although we may not see the causes, the effects will be there. An integrated view of oneself, one's relationship with God and an understanding of the whole of creation is, I believe, most important for each one of us to work out for himself. Such a wholistic view is certainly not to be obtained in most public schools today, nor taught in popular science books found in the library! In this chapter, I would like, therefore, to return to the realm of the spiritual and summarize very briefly what the Bible says about the principal inhabitants of the spiritual world; that is, the angels.[1]

## Cherubim and Seraphim

In the hierarchy of the angels as revealed to us in Scripture, the greatest and mightiest are known as the "cherubim," among whom are the "Four Living Creatures." The King James Version of the Bible calls them "the four beasts," but this is a poor translation of the Greek, *zoon*, meaning living beings. They are remarkable creatures first described for us in detail around the central throne of God by the prophet Ezekiel at the time of the Babylonian captivity of the Jews in July 593 B.C.

Isaiah may have glimpsed them earlier in his vision recorded in Isaiah 6. The Hebrew word found in the latter passage is, however, Seraphim, ("burning ones"), rather than Cherubim, so many Bible scholars treat the Seraphim as a separate, but very high class of angels. Our common notion of "cherubs," as childlike harmless angels with tiny fluttering wings is utterly unbiblical!

Cherubim with whirling, flaming swords guarded the gate of access back into Eden after Adam and Eve were expelled, to keep them from eating of the Tree of Life in their fallen condition. That is, they could not return to fellowship with God except by the way of the cross. Two figures of cherubim with outstretched arms covered the mercy seat which sat on the top of the Ark of the Covenant in the Tabernacle of Moses and later in the Temple of Solomon.

## Visions, Dreams, and Revelations

Ezekiel had the clearest visions of God seen by all the prophets. As a young man in his twenties, he was carried into captivity to Babylon (along with Daniel) but was allowed to see the glory and splendor of the Holy One of Israel. He was given to know details about the long term future of the nation and to foresee the last and greatest temple the Jews would one would one day build under Messiah's personal direction. Visions differ from dreams and hallucinations in the Biblical record. God sometimes spoke audibly to individuals in the Bible; at other times, He influenced men such as Joseph and Daniel by giving clear and specific guidance through dreams.

Today, now that Scripture is complete, God's "last word" to man is Jesus, as the opening verses of the letter to the Hebrews make clear. A clear warning is also in order - not all visions come from God. Hallucinations, such as those experienced by individuals on drugs, or under severe emotional stress, or deprived of sleep, are usually heightened imaginations. Many false religions and cults base their teachings on the visionary experiences of their founders. Islam was born because the "prophet" Mohammed received a "vision" from an angel. The Mormon Cult was initiated by a supposed "divine revelation" given to Joseph Smith. Demons and evil spirits and Satan himself coming "as an angel of light" are capable of counterfeiting experiences so that they appear to be coming from God when they are not. Such visions have most assuredly not come from the God who gave us the Bible.

Genuine visions, such as those seen by Ezekiel, Daniel, the Apostle John and other Biblical writers were experiences when the seer was allowed to briefly look behind the veil into the invisible world, ("the heavenly places"), into that more permanent and more solid world of the Spirit. Today, only if an experience can be confirmed by direct words from scripture can the believer be sure what is happening is coming from God. The Christian faith is not existential, not based on experience or emotion, but grounded in a thorough and complete body of knowledge given to us in the Bible. Biblical events are historically true as well, not at all based upon myth, folklore or tradition which form the backdrop of common world religions.

## Ezekiel's Vision

In regard to the Four Living Creatures, Ezekiel records,

"As I looked, behold, a storm wind came out of the north, and a great cloud, with brightness around about it, and fire flashing forth continually, and in the midst of the fire, as it were gleaming bronze. And from the midst of it came the likeness of four living creatures. And this was their appearance: they had the form of men, but each had four faces, and each had four wings. Their legs were straight, and the soles of their feet were like the sole of a

173

calves foot; and they sparkled like burnished bronze. Under their wings on their four sides they had human hands. And the four had their faces and their wings thus: their wings touched one another; they went everyone straight forward without turning as they went. As for the likeness of their faces, each had the face of a man in front; the four had the face of a lion on the right side; the four had the face of an ox on the left side, and the four had the face of an eagle at the back. Such were there faces. And their wings were spread out above; each creature had two wings, each of which touched the wings of another, while two covered their bodies. And each went straight forward, wherever the spirit would go they went, without turning as they went.

"In the midst of the living creatures there was something that looked like burning coals of fire, like torches moving to and fro among the living creatures; and the fire was bright, and out of the fire went forth lightning. And the living creatures darted to and fro like a flash of lightning. Now as I looked at the living creatures, I saw a wheel upon the earth beside the living creatures, one for each of the four of them. As for the appearance of the wheels and their construction: their appearance was like the gleaming of a chrysolite; and the four had the same likeness, their construction being as it were a wheel within a wheel. When these went they went in any of their four directions without turning as they went. The four wheels had rims and they had spokes; and their rims were full of eyes round about. And when the living creatures went, the wheels went beside them; and when the living creatures rose from the earth, the wheels rose. Wherever the spirit would go, they went, and the wheels rose along with them; for the spirit of the living creatures was in the wheels. When those went, these went, and when those stood, these stood; and when those rose from the earth, the wheels rose along with them, for the spirit of the living creatures was in the wheels.

"Over the heads of the living creatures there was the likeness of a firmament (the "sea of glass", probably metallic gold-glass) shining like crystal, spread out over the heads. And under the firmament their wings were stretched out straight, one toward another; and each creature had two wings covering its body. And when they went, I heard the sound of their wings like the sound of many waters, like the thunder of the Almighty, a sound of tumult like the sound of a host; when they stood still they let down their wings. And their came a voice from the firmament over their heads; when they stood still, they let down their wings. And above the firmament over their heads there was the likeness of a throne, in appearance like sapphire, and seated about the likeness of a throne was as it were the likeness of a human form. And upward from what had the appearance of his loins I saw as it were gleaming bronze, like the appearance of fire round about; and downward from what had the appearance of his loins I saw as I saw as it were the appearance of fire, and there was brightness around about him. Like the appearance of the bow that is in the cloud on the day of rain, so was the appearance of the brightness round about. Such was the appearance of the likeness of the glory of the Lord." (Ezekiel 1:4-28, see also Chapter 10) (Modified KJV).

## Four Mighty Angels Around the Throne

Numerous Bible scholars have commented upon the fact that the Four Living Creatures, the Cherubim, correspond one-to-one with the four Gospels with which the New Testament opens. Lion, ox, man, and eagle all symbolize various attributes of God in four-fold symmetry. Since the number "four" is symbolic of the world in scripture, we have in the four cherubim around the throne, a picture of God's government and concern for all aspects of the physical world, and the world-system which we call "human society." The lion corresponds to the Gospel of Matthew - here Jesus is seen as the Lion out of the tribe of Judah, the rightful king, depicting God's kingly sovereignty over Israel and the nations. The ox pictures for us servitude and sacrifice (Mark); The Cherubim's face, like that of a man, corresponds to Luke's Gospel and pictures divine intelligence; and the eagle symbolizes the sovereignty and deity of God (John's Gospel). The four Gospels are also God's complete message to the descendants of Shem, Ham, Japheth, and to mankind in general, in the same order.[2] The "wheels" of Ezekiel appear to symbolize divine mobility, (the ability of God to observe, to move and to act, anywhere and at anytime, unrestricted by the limitations of the material world). The entire vision seems to picture for us God's government of human affairs and of nature.

Similarly, the cherubim, the highest class of the angels, are also described for us in detail by John the Apostle on the Isle of Patmos about 95 AD:

"At once I (John) was in the Spirit, and lo, a throne stood in heaven, with one seated on the throne! And he who sat there

175

appeared like jasper and carnelian, and round the throne was a rainbow that looked like an emerald. Round the throne were twenty-four thrones, and seated on the thrones were twenty-four elders, clad in white garments, with golden crowns upon their heads. From the throne issue flashes of lightning, and voices and peals of thunder, and before the throne burn seven torches of fire, which are the seven spirits of God; and before the throne there is as it were a sea of glass, like crystal. And round the throne on each side of the throne, are four living creatures, full of eyes in front and behind (depicting the omniscience of God): the first living creature like a lion, the second living creature like an ox, the third living creature with the face of a man, and the fourth living creature like a flying eagle. And the four living creatures, each of them with six wings, are full of eyes all around and within, and day and night they never cease to sing, 'Holy, holy, holy is the Lord God Almighty, who was and is and is to come!' And whenever the living creatures give glory and honor and thanks to him who is seated on the throne, who lives for ever and ever, the twenty-four elders fall down before him who is seated on the throne and worship him who lives for ever and ever; they cast their crowns before the throne, singing, 'Worthy are thou, our Lord and God, to receive glory and honor and power, for thou didst create all things, and by thy will they existed and were created'." (Revelation 4:2-11)

## Angels and the Government of the Universe

The Cherubim also seem to represent God's Intelligence Agency - with their all-seeing eyes and ability to move quickly in all four directions. The number "four" indicates they have a central role in God's government of human society. The Four Living Creatures represent the topmost level of the system of angels who fall into definite hierarchy and class structures hinted at in scripture. For example, Paul says there are angelic principalities (*arche*), powers (*exousia*), thrones (*thronoi*), dominions (*kuriotes*), and powers (*dunamis*). Thus, there are angels with power over a nation or a city or a planet or a star such as our sun. In addition to all manner of greater and lesser angels, the Mighty Archangel Michael, guard of Israel, and the great messenger Gabriel are also named. (Only Michael, of all the angels, is actually called an archangel in scripture).

God created great numbers of spirits, some now fallen and known now as demons. Demonic forces operate in all forms of idolatry and false religion, for instance. Demons, deceiving spirits, familiar spirits, and unclean spirits do not have bodies of their own; hence, they seek to possess humans (and can possess animals also). The popular New Age fad of "channelling" appears to be a good example of demons acting through men and women who make themselves available to evil spirits so as to receive hidden, esoteric wisdom alleged to come from the past or from "other worlds."

Guardian angels watching over the lives of God's people are mentioned in the opening verses of the letter to the Hebrews, in the Psalms and in the Gospels. In the Book of the Revelation we are told of four angels who have charge over the "four winds that blow from the four corners of the earth." We also read of four fallen, now-bound, angels who are to be unloosed at the Euphrates River. The Greek word *angelos* and the Hebrew word *malakh* mean "messenger," but it is clear in the Bible that angels do much more than run errands and deliver telegrams for God.

Great and mighty angels will take charge in the unfolding of God's terrible judgments on the earth that are to come upon us at the end of the present age as described in the last book of the Bible. As mentioned earlier, angels seem to have control over the forces of nature and human events in their roles as instruments of God. This does not cause man to be a mere puppet, tossed, and buffeted about by the unseen activity of these mighty beings. However, the Bible emphasizes that we cannot understand what is happening in the world or in history without understanding the activity of God and of the angels.

## *The* Angel of the LORD

One special angel is "*The* Angel of the Lord" who makes a number of appearances in the Old Testament. These are believed by careful Bible scholars to be a theophanies, or appearances of the Son of God, prior to His incarnation in Bethlehem:[4]

### An Angel You Ought to Know
"The modern mind cannot conceive of angelic beings. This is due in part to medieval art and literature, which relegate belief in angels to the realm of superstition. Or perhaps we like to try to

explain away that which makes us uncomfortable. Indeed, there are those who would even dismiss the belief in God as mere superstition.

"Yet virtually every philosopher who has recognized the God of the Bible has also believed in angels - not the cute cherubs of Christmas cards, but mighty and powerful spiritual beings who are servants of the Most High God.

"In Hebrew, the word for "angel" is *malakh*. A *malakh* is a messenger, either human or angelic. Yet there is one *malakh* who stands out from all the rest. The Bible calls him simply, "the angel of the Lord."

"Since the time of Abraham, our people have known about the angels of the Lord. In the Talmud he is given the name Metatarsus, which indicates a special relationship with God. One meaning of *meta* and *thronos*, two Greek words, gives the sense of "one who serves behind the throne of God." He is also known as the "Prince of the Countenance" because of the close proximity between this angel and God Himself. The implication for the *malakh* of the Lord is that He is, above all, the messenger of God, the one sent by God, the one who represents God Himself.

"Throughout the Tenach, the angel of the Lord often appeared in human form. He served in three ways - guiding the people of Israel, effecting miracles, and executing judgment on Israel's enemies.

"He is first mentioned in Genesis 16. After Hagar fled into the wilderness to escape from Sarah, Abraham's wife, the angel of the Lord found her and admonished her to return to her mistress. He then promised to greatly multiply her descendants and prophesied the birth of Ishmael, who became the progenitor on the Arab nations.

"In Genesis 22, read every Yom Kippur, it is the angel of the Lord who called from heaven to stay the hand of Abraham as he took the knife to slay his son Isaac. In Exodus 14, he was in the pillar of cloud guiding the Israelites through the wilderness after their flight from Egypt. In Numbers 22:22-35, the angel of the Lord appeared to Balaam, the non-Jewish prophet, and gave him orders to be followed.

"He instructed Gideon in Judges 6, telling him to deliver Israel from Midian. He prophesied the birth of Samson (Judges 13), directed Elijah to Mt. Horeb (I Kings 19), and commanded King David to build an altar in Jerusalem which later became the site of the temple of Solomon (I Chronicles 21:18).

"The angel of the Lord is also presented to us as an avenger of evil, a judge. When Assyria, which was one of the ancient super powers, threatened to destroy Israel (700's B.C.E.), it was the angel of the Lord who killed the 185,000 Assyrian soldiers besieging Jerusalem (2 Kings 19:35). This angel, powerful in battle, was gentle enough to succor a fleeing and frightened Hagar in the wilderness.

"This angel was perceived in a unique and remarkable way by those with whom he came in contact. In ancient times it was common knowledge that if one saw God, it meant death for the individual. God stated this directly to Moses on Mt. Sinai: "You cannot see my face, for no one may see me and live" (Exodus 33:20). After Hagar saw the angel of the Lord, it is recorded that she called him *Lord* and marveled that she was still alive after having seen him (Genesis 16:33).

"Jacob reacted in a similar fashion when he wrestled with a "man" during the night. The man blessed Jacob and changed the patriarch's name to Israel. Jacob responded by calling the place of this encounter Peniel, saying, 'it is because I saw God face to face, and yet my life was spared.' (Genesis 32:30). Jacob identified the "man" as God. Later in his life, when Jacob blessed his son Joseph and his children, he said, "The God before whom my fathers Abraham and Isaac walked, the God who has been my Shepherd all my life to this day, the Angel who has delivered me from all harm..." (Genesis 48:15,16). The parents of Samson, likewise, recognized the angels of the Lord to be God, "We are doomed to die!..We have seen God!" (Judges 13:22).

"The angel of the Lord appeared to Moses in the midst of a burning bush (Exodus 3:2) but then in verse 4, "*God* called to him from with the bush..." When the Lord delivered the children of Israel from Egypt, the Bible says, "By day the Lord went ahead of them in a pillar of cloud to guide them on their way and by night in a pillar of fire to give them light..." (Exodus 13:21). But we read again in Chapter 14, verse 19, that the "angel of God, who had been travelling in front of Israel's army, withdrew and went behind them. The pillar of cloud also moved from in front and stood behind them, coming in between the armies of Egypt and Israel" (Exodus 14:19,20). And then in verse 24 we are told that the Lord looked down on the Egyptian army through the pillar of fire and cloud, and fought against Egypt! Who is involved in this pillar - the angel of the Lord or God Himself?

"In Judges 6, the angel of the Lord appeared to a timid Gideon and sat down under an oak tree to initiate a conversation with him (vss. 11,12). In verse 13, we see Gideon responding, but in verse 14 something strange happens: all of a sudden it is the *Lord* who is seen talking to Gideon! In verse 16, the conversation continues, but in verse 20, it is the angel of God who is in conversation. The next verse relates a miracle is performed by the angel. Then Gideon responds: 'Ah, Sovereign LORD! I have seen the angel of the LORD face to face!' But the LORD said to him, 'Peace! Do not be afraid. You are not going to die!'" (Judges 6:22,23)

"Are there two or three characters in this passage? One, of course, is Gideon. In verses 11 and 12 we have the angel of the Lord, then the Lord in verses 14 and 16, then the angel of God in verse 20 and again the angel of the Lord in verse 21. This writer maintains that the angel of the Lord must be the Lord God. Yet in some sense, the angel of the Lord, even though He Himself is deity, must be distinguished from the totality of the Godhead. For in Zechariah 1:12, the angel of the Lord is seen interceding on behalf of Israel, calling out to the Lord of hosts! The Holy Scriptures have given us a paradox: The Angel of the LORD is distinct from God, yet is Himself very God!

"This paradox is consistent with God's very nature. God, who is involved with His creation and interested in our welfare (Psalm 139:3, 13) is also high above (Isaiah 55:8,9). God is a vengeful God to those who flaunt His revealed will (Deuteronomy 32:35), and yet He is merciful (Exodus 33:19). God is all-important (Psalm 139), and yet He willingly "forgets" (Jeremiah 31:34, Isaiah 64:9). God is an advocate, a defender of His people (Psalm 59:1, Job 16:19), but He is also a prosecutor and judge (Psalm 50:6), Ecclesiastes 3:17). When we study the nature of God, we find paradoxes.

"The angel of the Lord, God Himself, revealed Himself in a visible, personal way - taking the form of a human being. This writer maintains that not only could the angel of the Lord *assume* human *form*, but that, in time, he *took on* true humanity by being *born* into the human race!

**'Who, being in very nature God, did not consider equality with God something to be grasped, but made himself nothing, taking the very nature of a servant, being made in human likeness."** (Philippians 2:6,7) (NIV)

"This writer also maintains that the Old and New Testaments are intrinsically connected and make up God's revelation to man. The claims in the New Testament portion concerning Jesus correspond to those claims in the Old Testament portion which refer to the angel of the Lord. Jesus claimed to be the supreme *malakh* of God. "Anyone who has seen me has seen the Father" (John 14:9). The angel of the Lord did miraculous acts; so did Jesus. (See John 2:9, Matthew 8:3, Luke 7:11, Matthew 15:32, etc.) The angel of the Lord taught and instructed people; Jesus was called "rabbi" (John 20:16). The angel of the Lord is a judge of mankind; in John 5:22 we see "The Father judges no one, but has entrusted all judgment to the Son." Is Jesus of Nazareth and the angel that wrestled with Jacob one and the same? Carefully study the Scriptures for God's answer."

## The Armies of the LORD

The "angelic host" is under the direction of "The Lord of Hosts" (*Yahweh Sabaoth*) or the "Lord of Armies." In the New Testament, Jesus heads up these military resources of the Godhead which include the angels, and eventually also the saints of God. The active warfare in heaven between good and evil in the spiritual realm involves conflicts between fallen and unfallen angels as well as campaigns having to do with angelic interactions with human affairs. An interesting incident in the life of Elisha the prophet (to give a behind-the-scenes glimpse at God's angelic armies) is given in 2 Kings 6. Alarmed at the ability of Elisha to warn the armies of Israel against impending attacks from the king of Syria, the latter attempts to capture Elisha when he was at Dothan. The sequence of events which follows clearly shows clearly the course of events is in reality controlled not by visible circumstances or "good luck," or by Elisha's cleverness, but by the Lord working through angels:

**"So he** (the king of Syria) **sent horses and chariots and a great army; and they came by night, and surrounded the city. When the servant of** (Elisha) **the man of God rose early in the morning and went out, behold, an army with horses and chariots was round about the city. And the servant said, 'Alas my master! What shall we do?' Elisha said, 'Fear not, for those who are with us are more than those who are with them. Then Elisha prayed, and said, 'O LORD, I pray thee, open his eyes that he may see.' So the LORD opened the eyes of the young man, and he saw; and behold the mountain was full of horses and chariots of fire round about**

Elisha. And when the Syrians came down against him, Elisha prayed to the LORD, and said, 'Strike this people, I pray thee with blindness.' So God struck them with blindness in accordance with the prayer of Elisha.

"And Elisha said to the Syrians, 'This is not the way, and this is not the city; follow me, and I will bring you to whom you seek.' And he led them to Samaria. As soon as they entered Samaria, Elisha said, 'O LORD, open the eyes of these men, that they may see.' So the LORD opened their eyes, and they saw' and lo, they were in the midst of Samaria. "When the king of Israel saw them he said to Elisha, 'My father, shall I slay them? Shall I slay them?' Elisha answered, 'You shall not slay them. Would you slay those whom you have taken captive with your sword and with your bow? Set bread and water before them; that they may eat and drink and go to their master.' So he prepared for them a great feast; and when they had eaten and drunk, he sent them away and they went to their master. And the Syrians came no more on raids into the land of Israel." (2 Kings 6:14-23) (Modified KJV)

## Angels at War with One Another

Another interesting illustration showing how angelic activity interrelates with human affairs is given for us in the book of Daniel. Prayers by Daniel had reached the ears of God immediately (faster than the speed of light), but the answer, sent by angelic messenger, was delayed due to a mighty conflict between the messenger angel and evil angels who had charge of the affairs of nations around Israel:

"In those days I Daniel was mourning three full weeks. I ate no delicacies, no meat nor wine entered my mouth, nor did I anoint myself at all, till three whole weeks were ended. And in the twenty fourth day of the first month, as I was by the side of the great river, that is the Tigris, I lifted up my eyes, and looked, and behold a man clothed in linen, whose loins were girded with fine gold of Uphaz: His body was like beryl, and his face like the appearance of lightning, and his eyes like flaming fire, his arms and his feet were like in color like burnished bronze, and the voice of his words like the noise of a multitude. And I, Daniel, alone saw the vision: for the men that were with me saw not the vision; but a great trembling fell upon them, and they fled to hide themselves. Therefore I was left alone, and saw this great vision, and there remained no strength in me: for my radiant appearance was fearfully changed, and I retained no strength. Yet heard I the voice of his words: and when I heard the voice of his words, then I fell on my face, in a deep sleep with my face toward the ground.

"And, behold, a hand touched me, and set me upon my hand and knees. And he said to me, 'O Daniel, man greatly beloved, understand the words that I speak to you, and stand up: for to you I have been sent.' And when he had spoken this word to me, I stood trembling. Then said he to me, 'Fear not, Daniel: for from the first day that you set your heart to understand, and to humble yourself before your God, your words were heard, and I have come because of your words. But the prince of the kingdom of Persia withstood me twenty one days: but, Michael, one of the chief princes, came to help me; and I remained there with the kings of Persia. Now I have come to make you understand what shall befall your people in the latter days: for the vision applies to man days from now' And when he had spoken these words to me, I set my face toward the ground, and I became dumb.

"And, behold, one in the likeness of the sons of men touched my lips: then I opened my mouth, and spoke, and said to him who stood before me, 'O my lord, by reason of the vision pains have come upon me, and I have retained no strength. How can the servant of my lord talk with my lord? For as for me, there remains no strength in me, nor is there breath left in me.' "Then one who had the appearance of a man came again and touched me, and strengthened me, and said, 'O man greatly beloved, fear not: peace be to you, be strong, yes, be strong.' And when he had spoken to me, I was strengthened and said, 'Let my lord speak; for you have strengthened me. Then said he, 'Do you know where I have come from to you? And now will I return to fight with the prince of Persia: and when I have gone forth, lo, the prince of Greece will come. But I will show you that which is noted in the scripture of truth: and there is no one who stands with me in these things, but Michael your prince." (Daniel 10:2-21) (Modified KJV)

Daniel was in Babylon at the time of the experience described above. The great angel (probably Gabriel) who came to assist him was at the time "fighting with the prince (angel) of Persia," and he mentioned that after that would come "the prince (angel) of Greece". Here, we get a hint that angelic activity then taking place in the heavenlies (involving God's angels warring against the angel over Persia) was relevant to current events in Babylon. Later in human history, the angel over Greece (in the time of Alexander the Great who came to power thereafter) would be in ascendancy.

## Angels Work Behind the Scenes of History

In Daniel, Chapter 12, the prophet records that the angel responsible for Israel, Michael, would "arise" at the close of the latter days on behalf of God's people Israel. Associated with the rise of Michael would come the throwing down to earth of Satan and the "time of Jacob's trouble" spoken of by the prophet Jeremiah, **"a time of trouble such as never has been since there was a nation till that time..."** (Daniel 12:1-3) Thus, angelic activity in heaven runs parallel with events on earth though it is difficult to establish a one-to-one correspondence between events transpiring in heaven and the timing of events on earth.

In Zechariah, Chapter 1, scripture records dreams of the prophet in which he saw horses riding to the corners of the earth. These horses symbolize divine activity in human affairs as do the more famous "Four Horsemen of the Apocalypse" (Revelation, Chapter 6). In the Book of the Revelation the attention of the reader is turned to events in heaven; then the scene switches back to events transpiring on earth in human history. As the Book of the Revelation unfolds, the two realms merge so that there is a closer connection ("closer coupling") between the two worlds. This is the meaning of "apocalypse" - "unveiling".

Angels keep records and execute judgments for God (Ezekiel 8-9). Angels, acting for God, reduced the mighty world-conqueror Nebuchadnezzar to seven years of psychotic wanderings to humble him so that he would come to see who was really running the universe (Daniel 4).

Angels announce important events in scripture, such as the birth of Jesus, for example. Ordinarily when angels appear among men, they take on human form and may even be mistaken for ordinary men. Hence, the writer of the letter to the Hebrews tells us, **"Do not neglect to show hospitality to strangers, for thereby some have entertained angels unawares."** (Hebrews 13:2) The writer no doubt had in mind the visit of *The* Angel of the LORD and two other angels to Abraham's tent on their way to

184

destroy Sodom and Gomorrah (Genesis 18). After a meal and a con-versation which included Abraham's petitions for the doomed cities, the two other angels departed and, upon arriving in Sodom, were treated by the men of that city as handsome, desirable young men with whom they wished to engage in immoral sexual activities. These angels rescued Lot who apparently seems to have remained in some doubt about their true identity the whole time. Stories of angels appearing occasionally among men in contemporary times seem to be well-founded in some instances.

In the Book of the Revelation, angels are frequently seen in dazzling appearances. The aged Apostle John, who had spent his teen age years travelling with Jesus as one of the apostles, and after that, six decades of godly ministry in daily fellowship with the living Lord, was nevertheless so overwhelmed by the splendor of the angels he saw in his visions that he was twice tempted to worship them thinking they were actually the Lord in His risen splendor! Demons and evil spirits, on the other hand, seem incapable in taking on physical appearance but gain access to the physical world by possessing men. Demon possession is not uncommon in the world today and occasionally encountered in the United States due to rapidly declining moral conditions and open Satan worship.

## Man Involved in the Angelic Conflict

The New Testament also emphasizes that the Christian cannot understand what is happening to him and to the world he lives in unless he takes the angelic conflict seriously. Although I mention just two passages, the subject deserves very careful study by every Christian. The Apostle Paul writes,

**"For though we live in the world (system) we are not carrying on a worldly war, for the weapons of our warfare are not worldly (carnal) but have divine power to destroy strongholds. We destroy arguments and every proud obstacle to the knowledge of God, and take every thought captive to obey Christ, being ready to punish every disobedience, when your obedience is complete."** (II Corinthians 10:3-6).

Paul also writes clearly on this subject in the closing portion of his letter to the church at Ephesus:

**"Finally, be strong in the Lord and in the strength of his might. Put on the whole armor of God, that you may be able to withstand**

against the wiles of the devil. For we are not contending against flesh and blood, but against the principalities, against the powers, against the world rulers of this present darkness, against the spiritual hosts of wickedness in the heavenly places. Therefore take the whole armor of God, that you may be able to withstand in the evil day, and having done all, to stand. Stand therefore, having girded your loins with truth, and having put on the breastplate of righteousness, and having shod your feet with the equipment of the gospel of peace; besides all these, taking the shield of faith, with which you can quench all the flaming darts of the evil one. And take the helmet of salvation, and the sword of the Spirit, which is the word of God. Pray at all times in the Spirit, with all prayer and supplication. To that end keep alert with all perseverance, making supplication for all the saints." (Ephesians 6:10-19)

## The Greatest of the Fallen Angels

The Bible does not explain or justify the existence of God or of the devil; the existence of evil is already clear by Genesis, Chapter 3. As the Bible unfolds, we glean more and more information about the Tempter. He first appears in Genesis as "The Shining One." Most Bible scholars find that on the surface Isaiah 14 and Ezekiel 28 passages seem to be written about evil earthly kings, but they are also figurative pictures telling us much more. Ultimately Satan becomes incarnate in the "man of sin" or the Antichrist, so it is not surprising that his attributes can be seen in powerful and evil human rulers of the past: the King of Babylon, the King of Tyre, the Caesars, Hitler, etc. Isaiah and Ezekiel seem to be telling us about one of the Mighty Cherubim, perhaps the greatest of the archangels, who fell because of pride:

"How are you fallen from heaven, O Lucifer, son of the morning! How are you cut down to the ground, you who laid the nations low! You said in your heart, 'I will ascend into heaven, I will exalt my throne above the stars of God: I will sit upon the mount of the assembly, in the far north: I will ascend above the heights of the clouds; I will be like the most High'. Yet you shall be brought down to Sheol, to the depths of the Pit. Those who see you shall stare at you, and ponder over you, 'Is this the man who made the earth tremble, who shook kingdoms; who made the world like a desert, and overthrew its cities; who did not let his prisoners go home? All the kings of the nations lie in glory, every one in his own tomb. But you are cast out of your sepulchre, like a loathed, untimely birth, clothed with the slain, those thrust through with a sword, who go down to the stones of the Pit; like a dead body trodden under feet you will not be joined with them in burial, because you have destroyed your land, and slain your people. 'May the descendants of evildoers nevermore be named! Prepare slaughter for his sons because of the guilt of their fathers, lest they rise and possess the earth, and fill the face of the world with cities.'" (Isaiah 14:12-22) (Paraphrased KJV)

From Ezekiel,

"Moreover the word of the Lord came to me, saying, 'Son of man, take up a lamentation over the king of Tyre, and say to him, 'Thus says the Lord God; you were the signet of perfection, full of wisdom, and perfect in beauty. You were in Eden the garden of God; every precious stone was your covering, carnelian, topaz, and the diamond, beryl, onyx, and jasper, sapphire, emerald, and lapis lazuli, and gold; and wrought in gold were your settings and your engravings. On the day that you were created they were prepared. "You were the anointed guardian cherub who covers (guards), and I placed you there. You were upon the holy mountain of God; you walked in the midst of the stones of fire. You were blameless in your ways from the day that you were created, until iniquity was found in you. In the abundance of your trade you were filled with violence, and you sinned: so I cast you as profane from the mountain of God. And I have destroyed you, O covering (guarding) cherub, from the midst of the stones of fire. Your heart was lifted up because of your beauty, you corrupted your wisdom because of your splendor: I cast you to the ground, I exposed you before kings, that they might feast their eyes on you. You have defiled your sanctuaries by the multitude of your iniquities, in the unrighteousness of your trade. Therefore I have brought forth fire from the midst of you, it consumed you, and I have turned you to ashes on the earth in the sight of all those who see you. All who know you among the peoples are appalled at you: you have come to a dreadful end, and shall be no more forever." (Ezekiel 28:1-19) (Paraphrased KJV)

187

## Lucifer, the "Light-Bearer"

"Lucifer" is the "light bearer" or very possibly *the angel who has stewardship over light*. As the "anointed guardian Cherub" in the garden of God (a spiritual place corresponding with the garden in Eden on earth), he may well have had enormous power and influence over the entire angelic host. In the New Testament, Paul warns that he sometimes comes among men disguised as an angel of light (presumably through false teachers and also as a giver of deceptive visions) though, in fact, he has now become the Prince of Darkness. Lucifer may have been originally God's appointed watch guard over the forces of nature; for example, he could have been in charge of such physical phenomena as the propagation of light. Thus one possibility raises itself: that the velocity of light began to decrease at the time Satan fell. I personally am intrigued by this possibility but can offer no proof from scripture to back this hypothesis.

It is certainly appropriate for us to re-examine the entire visible, created universe and look for evidences of the destructive work of Satan since his fall, for there is no doubt from scripture that he has wrought havoc in man and in nature since he turned away from serving God. It is also possible the observed decrease in the velocity of light began with the fall of man. The third possibility is that c-decay is a natural effect built into the universe by God from the start.

## That Ancient Serpent

Satan is given a number of titles and descriptors in the Bible. He is "the prince of this world (*cosmos*, the world system), "the prince of the power of the air," "the god of this age (*aion*), "the prince of demons", "the devil", "that ancient serpent", "the great red dragon", "the evil one", "the destroyer", "the tempter", "the accuser of the brethren", "the spirit that now works in the sons of disobedience", "a liar and a murderer from the beginning," "*the* deceiver". His actual power over the world can be seen in the offers he made to Jesus during the latter's 40 days' temptation in the wilderness, recorded in Matthew, Chapter 4. Satan is the ruler over fallen angels, believed to number one-third of the angelic host according to Revelation 12:4. He is the opposer of God's Person; the counterfeiter of truth, "the deceiver of nations".

Everything Satan does he must do by direct permission of God, as the book of Job clearly shows, so it is not as if there were two warring gods, one good and the other evil, in the universe. The devil is a defeated enemy, his evil plan and power were undone by the work of Jesus on the cross though we wait in our time for his final defeat and removal from the heavens and the earth, by Michael the Archangel. If we look at circumstances around us it seems he is winning and gaining ground unchecked. If we open our Bibles, however, his defeat is spoken of repeatedly as an already accomplished fact. Christians have not need to fear the devil, although we are warned by Peter,

**"Be sober, be watchful. Your adversary the devil goes around like a roaring lion, seeking some one to devour. Resist him, firm in your faith, knowing that the same experience of suffering is required of your brotherhood throughout the world. And after you have suffered a little while, the God of all grace, who has called you to his eternal glory in Christ, will himself restore, establish, and strengthen you."** (I Peter 5:8-10)

These notes on the angels have been very brief and the reader is strongly urged to become well acquainted with all of Scripture and to read classic Christian books on spiritual warfare and angelology to gain a better understanding of the role of good and fallen angels in creation. Difficult though it is for us to grasp, the laws of physics are apparently not operative apart from the angels who are part of the management system of God's government of creation. Just how angelic activity affects events in the physical world is not known. Actual transfer of new energy may not necessarily be involved most of the time when angels act since such energy transfers would appear to us to be violations of the laws of conservation of energy which as I have said do not appear to have been violated in any known instance observed by men.

## Angels and Planetary Wars

In a computer conference held a few years ago between a group of scientists across the United States concerning the interesting geological history and some unusual surface features of the planet Mars,[5] a colleague of mine suggested that perhaps Mars had once been attacked by fallen angels and that the purpose of the hypothesized angelic war involving Mars and other planets, might have been to prevent the birth of the Savior of mankind whose coming was foretold in the Bible as early as Genesis, Chapter 3. The earth has been central to God's plan for the redemption of mankind. On no other planet did Christ become incarnate, and no where else did He die. It is not inconceivable, therefore, that the other planets were once angelic "watchposts" guarding the earth. Satanic attempts to destroy the earth through angelic warfare following his fall could have reduced all the planets (except earth), to ruined, desolate, barren worlds. This is an unusual speculation but would serve to illustrate yet another of the devil's attempts to prevent the redemption of man. From scripture we do know that good and fallen angels war against one another, evidently using powerful weapons rather than mere swords of steel. We would expect such mighty creatures, with such vast influence, to have weaponry in advance of that imagined by our government's SDI programs, and beyond the conceptions raised for our entertainment in the "Star Trek" and "Star Wars" series.

## Evil Seeks to Thwart the Work of God and His Messiah

The curse on the serpent in Genesis is described as perpetual enmity between the godly line of Eve and the enemy of mankind, Satan: "**I will put enmity between you and the woman** (God says to the serpent) **and between your seed** (the Antichrist) **and her seed** (the Messiah); **he shall bruise your head and you shall bruise his heel.**" (Genesis 3:15) After Cain murdered Abel, another son, Seth, was born to Eve. This begins the so-called godly line of promise leading to Jesus. Matthew's Gospel, Chapter One, picks up this genealogical line from Abraham to Joseph showing that Jesus was the rightful legal heir to the throne of Israel on Joseph's side, but rightful blood heir on Mary's side also. Luke's Gospel, which records the line of Mary goes all the way back to Seth.

A fascinating Old Testament study of the "line of promise" clearly shows that Satan made numerous attempts to permanently interrupt this line so that the qualified Messiah, Jesus, would never be born and thus no legitimate heir would ever come to Israel and mankind. Reading the Old Testament accounts that center around the godly line of ancestors of Jesus shows clearly that Satan made many attempts, some almost successful, to destroy or interrupt the coming of the promised seed of the woman. To cite one example, Herod the Great (d. 4 B.C.), upon hearing of

190

the birth of Jesus from the wise men of the east, murdered all the male infants in Bethlehem under 2 years of age (Luke, Chapter 2); however, Joseph having been warned by an angel had already fled into Egypt with Mary and Jesus, where they remained for several years in a the part of Old Cairo still known today.

Even during the life of Jesus, the devil attempted numerous times to thwart Jesus' primary purpose of becoming a cosmic sacrifice for the sins of all the mankind. In addition to rescuing lost men, many theologians believe that Jesus also "ransomed" lost property and territory belonging to God from any claim the enemy may have had upon it, whether those claims were all legitimate or not. The Temptation of Jesus by the devil recorded in Matthew, Chapter 4 is illustrative of the promises Satan made to Jesus if the latter would but renounce his dependence upon the Father. If the four Gospels are read with this theme in mind, the enemy's desperate stratagems can be seen on every page. Fortunately for us, all were unsuccessful!

## War in Heaven and Satan Cast Down to Earth!

Satan's last attempt to rule the universe and overthrow God and His kingdom occurs on earth, not Mars. Scripture speaks of "many antichrists" who oppose Jesus and counterfeit His claims, but there is *the* Antichrist, the "man of sin" who will one day take charge of Western civilization. As part of His plan, God has determined to allow evil to run its full course on earth. Antichrist is to be an incarnation of Satan so that he, the Evil One, possesses, indwells and controls a man. Readers unfamiliar with the relevant scriptures should read Matthew 24-25, II Thessalonians 2, and Revelation 13, as well as the Book of Daniel.

Revelation 12 includes a description of the throwing down of the devil to earth in a yet-future time which will result in the greatest conflagration our planet has ever seen. This last war, beginning in space, is fought by the archangel Michael and his thousands of unfallen angels:

**'Now war arose in heaven, Michael and his angels fighting against the dragon; and the dragon and his angels fought, but they were defeated and there was no longer any place for them in heaven. And the great dragon was thrown down, that ancient serpent, who is called the Devil and Satan, the deceiver of the whole world - he was thrown down to the earth, and his angels**

were thrown down with him. And I (John) heard a loud voice in heaven, saying, 'Now the salvation and the power and the kingdom of our God and the authority of his Christ (Messiah) have come, for the accuser of the brethren has been thrown down, who accuses them day and night before our God. And they have conquered him by the blood of the Lamb and by their testimony, for they loved not their lives even unto death. Rejoice then, O heaven and you that dwell therein! But woe to you, O earth and sea, for the devil has come down to you in great wrath, because he knows his time is short!'" (Revelation 12:7-12)

## A Vision of Israel's History and Destiny

Revelation, Chapter 12 also contains a vision which has significance in the light of ancient, uncorrupted Hebrew astrology.[6] The Apostle John writes,

"And a great portent appeared in heaven, a woman clothed with the sun, with the moon under her feet, and on her head a crown of twelve stars; she was with child and she cried out in her pangs of birth, in anguish for delivery. And another portent appeared in heaven; behold a great red dragon, with seven heads and ten horns, and seven diadems upon his heads. His tail swept down a third of the stars of heaven, and cast them to the earth. And the dragon stood before the woman who was about to bear a child, that he might devour her child when she brought it forth; she brought forth a male child, one who is to rule the nations with a rod of iron, but her child was caught up to God and to his throne..." (Revelation 12:1-5)

Like many visionary, apocalyptic images in the Bible this picture is "outside of time". The woman is Israel and the child is Jesus. This passage of scripture suggests to us that one-third of the angels fell along with Satan and became instruments of evil. After depicting the 2000-year history of ancient Israel as a woman crowned with 12 stars representing the 12 tribes, the vision moves swiftly from the birth of Jesus to His ascension into the heavenly places 40 days after His resurrection. The passage then continues to describe the plight of the woman in the end-times (yet to come) when the small faithful remnant of believing Israel is forced to flee to Petra in Jordan to survive the ravages of a Great World War which will center in Israel. There, in the wilderness east and south of Jerusalem, Jesus will return to this faithful remnant, then appear in power and glory to conclude the battle of Armageddon. Isaiah 63 describes this return of Jesus first to Edom and Revelation 19:11ff, His coming in open splendor before the entire world to save a remnant of mankind from total self-destruction and to restore our broken planet.

## The Last War: Men Against God

The Bible has much to say about this final war. It begins with an invasion of the Holy Land by Egypt (Daniel 11: 36-45) and is followed by a full invasion by Russia (Ezekiel 38, 39). Other Old Testament prophets such as Joel, Isaiah, Zechariah, and Zephaniah add to our understanding of these events. The Olivet Discourse by Jesus (Matthew 24, 25) concentrates on these final events as they will be experienced by the Jews living in Israel at the time of the end when the Third Jewish Temple has been built, dedicated, and is functioning. The Russian threat to Israel is met by the Western powers: sending landing forces into Israel, probably at the sea port of Haifa, and probably through the use of nuclear weapons also. Russia is destined to be destroyed, however, by direct intervention of God in which nuclear warfare seems to play a part because **"fire rains down on the coastlands"** (the major continents of the Western world) according to Ezekiel. This is very likely a description of a limited nuclear war involving the destruction of European and American cities, as well as Moscow.

## The Coming Cosmic Invasion of Jesus

The early campaigns of Armageddon will pit the armies of earth against one another. A great horde of troops from China, Japan and the Far East will evidently join in, according to the descriptions given us in the Book of Revelation. In the later stages (coinciding with the throwing down of Satan from space to earth by Michael), the war between men on earth will apparently become a united attempt to thwart the approaching space armies of Jesus; that is, the church and armies of the "heavenly hosts," the legions of unfallen angels. Viewed from earth, the return of Jesus will no doubt be seen as an invasion from outer space by the last generation of evil men who have been fully and successfully deceived by Satan. God is fallen man's ultimate enemy whom they hate more than anyone else! Another illustration of the different timing of events as seen in heaven compared to earth is seen in Luke 10:18; Jesus said, "I saw Satan fall like lightning from heaven ". This is a prophetic passage from our point of view, but Jesus saw it as happening "now" in His frame of reference. As the "Great I AM", Jesus made a number of statements in the Gospel of John linking His oneness with Yahweh and His (Jesus') unique relationship with the Father. In several instances, His perception that He was fully aware of eternity (as none of us are) while He was on earth are evident from the things He said.

## Planets Devasted by Angelic Wars?

Perhaps there is some reason to suppose that Mars was the site of an ancient angelic "war" which was part of a great Satanic plan to destroy the line of Eve that would lead to Messiah and to establish his (Satan's) own total and final dominion over mankind. One of the mysteries in the geophysics of the planets is that most cratering on the moon, Mercury, Venus, and Mars seems to have occurred *early* in the history of these bodies. "Early bombardment" seems to have decreased greatly after about a billion years or so of radio-clock time had elapsed. If there were an angelic war on Mars long ago (and this we have not proven, only speculated about), it seems to us that it was not limited to Mars, but quite likely involved Venus and at least Mercury and the moon as well. Some have even wondered if the asteroid belt beyond Mars is indeed all that is left of a planet destroyed in an ancient war. A possible modified model of the solar system is this: All the planets (not just earth) were created "good"; that is, in pristine, "inhabitable" condition, but all except earth have been drastically altered by the angelic wars of aeons past. According to the Setterfield model for the decrease in speed of light, the most ancient events (appearing to be several billions of years in the past in atomic time) actually occurred only thousands of years ago in dynamical time. Angelic warfare has never ceased since the fall of Lucifer, and angelic wars are not over yet. This "threat from the heavens" is still with us, so we not only have the dim, ancient past to worry about; we have our own future to be ready for.[7]

Even though the hypothetical "war on Mars" (that is, an early geological disaster) appears to have occurred millions of years ago, this is radioactive clock time. The true age of the universe is apparently less than 8000 years so the radical changes on Mars could have been initiated by the onset of radioactive decay rates at the fall of man, or by collision with a comet, or by events connected in some way with the flood of Noah. Mars gives evidence of a ruined, wasted place (as do Venus and Mercury as well). Apparently Mars and Venus once had oceans - Mars has now lost its atmosphere and mild climate and its water is now frozen beneath the surface as ice or permafrost. Venus now has a dense atmosphere rich in sulfuric acid, active volcanoes, and a surface temperature comparable to that of molten lead! It is thus entirely possible that the solar system is not in its original created condition, but has been badly damaged by the activity of evil angels since the fall of Satan. Likewise, it is possible that many astronomical events visible today may not be events that are part of the original creation, but may be in reality destructive disruptions amongst the stars caused by the activity of the Evil One.

In several of the Old Testament battles (involving angels) in which God is said to participate directly, we find descriptions such as the throwing down of great hailstones from heaven and rains of fire and brimstone (sulfur), from heaven. This type of warfare is what one would expect on the planets if indeed they have been devastated and wasted by such a galactic war in the ages past. None of this speculation about ancient space wars can be "proven" from the Bible nor at present from scientific evidence but at least the Bible does not exclude such a possibility.

## The Shekinah Glory Returns

Ezekiel saw the shining presence of the Glory of God, the Shekinah (that is, "The Cloud") depart from the Temple in Jerusalem just prior to its destruction. This glorious cloud, into which Jesus ascended from the Mount of Olives forty days after His resurrection, will come again, for scripture says His splendrous presence will return to open visibility before all men. After Messiah's return, the cloud will remain over the Temple Mount in Jerusalem, according to Isaiah.

On the basis of what is said in scripture, it is quite likely that more frequent manifestations of angels will be common among men in coming days as our world continues to collapse and the heavenly places break through into our realm of space and time. For example, the Book of the Revelation begins with a clear distinction between "things on earth" and "things in heaven"; but as the Revelation unfolds the invisible is made known to men on earth. Those days will not be a time for fear, if one's life is entrusted into the hand of the Great Shepherd, but they will be terrible indeed for all those who continue in their unbelief as the unseen breaks in upon mankind and visionary experiences previously-rare will be everyday experience.

Concerning the end of our age, the prophet Joel has written,

**"And it shall come to pass afterward, that I will pour out my spirit on all flesh; your sons and your daughters shall prophesy, your old men shall dream dreams, and your young men shall see visions. Even upon the menservants and the maidservants, in those days, I will pour out my Spirit. And I will give portents in the heavens and on the earth, blood and fire and columns of smoke. The sun shall be turned to darkness and the moon to blood, before the great and terrible day of the Lord comes. And it shall come to pass that all who call upon the name of the Lord shall be delivered..."** (Joel 2:28-30) (Modified KJV)

●

## Notes to Chapter Nine

1. Almost any book on conservative Christian theology contains a section on Angelology. The leading Bible encyclopedias are quite helpful. A recent book I can recommend is C. Fred Dickason, *Angels, Elect and Evil* (Moody Press; Chicago 1975). Mystical and occult book stores carry books on angels that are usually misleading in that they go well beyond the bounds of Biblical revelation and in some cases are misleading. The Apocryphal Books of the Bible (found between the Old and New Testaments of some Bibles) name a few of the angels not mentioned in the inspired portions of scripture; however, I do not recommend these sources as especially helpful in understanding angels.

2. Arthur C. Custance has written an outstanding book on this subject, *Noah's Three Sons* (Zondervan Publishing; Grand

Rapids, 1975). The Bible student who wishes to study the family tree of the descendants of Adam and Eve and the line leading to Jesus the Messiah can obtain an excellent colored wall chart from the Good Things Company; Drawer N; Norman, Oklahoma, 73070. I have found this beautiful chart helpful again and again in studying Old Testament genealogies.

4. Concerning *The* Angel of the LORD, the following article, copyright 1980, is quoted in full, by permission of Jews for Jesus; 60 Haight Street; San Francisco, CA 94102. The author is Loren Jacobs.

5. See Randolfo Rafael Pozos, *The Face of Mars: Evidence for a Lost Civilization?* (Chicago Review Press; Chicago, 1986) and Richard C. Hoagland, *The Monuments of Mars: A City on the Edge of Forever* (North Atlantic Book; Berkeley, 1987).

6. Two outstanding older studies have been reissued because of their lasting value: E. W. Bullinger, *The Witness of the Stars* (Kregel Publications; Grand Rapids, 1893/1981) and Joseph R. Seiss, *The Gospel in the Stars* (Kregel Publications; Grand Rapids, 1882/1982).

7. C. S. Lewis has written a marvelous science fiction trilogy, *Out of the Silent Planet*, *Perelandra*, and *That Hideous Strength* which should be read by all Christians interested in the creation as a realm of men and of angels. The first book of the series concerns Mars which is found to have been partially devastated by the activity of the Evil One. Venus, in the second book is as yet an unfallen world, but the Tempter is actively trying to bring its downfall. Earth is the place of greatest darkness and the final battle ground between good and evil, in the third book of this trilogy. This last book is especially close to the Biblical picture concerning Earth as the center of the universe because it is the place of God's "universally effective" redemptive activity and the final battleground where Satan is ultimately totally defeated. Even though fictional accounts, C. S. Lewis is well-known in our time for his profoundly-accurate perceptions of Biblical reality. On the subject of spiritual warfare, I highly recommend Ray C. Stedman's *Spiritual Warfare: Winning the Daily Battle With Satan* (Multnomah Press; Portland, 1975).

# Chapter Ten

## Physics in a Fallen Universe

I have read many text books in physics, astronomy, and geology and even more Bible commentaries and treatises by creationists. Few writers in either group take an integrated, wholistic approach such as I am attempting in this present book. I hope the reader will be helped, rather than hindered, therefore, by my style of developing a topic from the Bible or from science, dropping it for a chapter or two, and then picking it up again. It has taken many years for me to be convinced about the things of which I have been writing. Even so, I am admittedly only scratching the surface and proposing some tentative new hypotheses in science which I hope will be provocative to students of the sciences and the Bible alike.

## God Created the Universe in A Series of Steps

Earlier I said I sometimes personally wished God had gone into more detail in giving us more information in the Bible on *how* He created the universe. The Lord chides Job for being truly ignorant of such matters (Job, Chapters 38-41), and whether we have a better understanding than Job is dubious! But given what we have, we can at least improve on existing purely secular models. For example, from Genesis, One we have seen that God created the universe out of nothing, that He made the raw material first and then formed and fashioned it into its final product. He did his work in six stages (days) and then ceased from further creative activity. God's creative activity took place over a fixed interval of time, and He created by means of a series of distinct events. This fact immediately rules out the Big Bang theory as having ultimate viability, for that theory extrapolates back from the present time epoch and does not take into account any of the known discontinuities in the laws of nature which can be inferred from the Bible.

Furthermore, physics is concerned with the physical world, which as we have seen, is not all there is to the universe. During each stage of creation, God introduced outside energy, created new matter, and introduced intelligent programming information to bring order out of chaos. The programming information God built into the old creation allowed the creation to continue without His constant ongoing interference, which would show up immediately as violations of the known laws of physics. (I do not find any evidence in science or the Bible which leads me to lend credibility to the view popularly-known as "theistic evolution.")

God's creative acts produced discontinuities in the laws of science when they occurred; and therefore, they are not easily treated mathematically even if we had been given more detail. In the creation of seeds, God gave all forms of life genetic codes sufficiently-complex to generate all the complexities and varieties of life man has ever seen or known. It is quite proper to say that each of us as individuals was "in Adam" at creation since all the genetic material we possess is derived from the original seed programmed into Adam.

God is a Spirit, and as we have seen, He operates from the spiritual realm of reality *into* the physical. As far as the physicist is concerned, God is an "outside Supreme Intelligence" who introduced both energy and programming information into the universe as it was built. In discussing miracles, I mentioned that I do not think that God regularly violates the present-day laws of physics; however, as the Sustainer of the universe, He is much more than a First Cause who built the universe and then left it to run unattended. That God may occasionally violate the laws of physics in working miracles, I have no way of proving or disproving from the Bible or from science. Many miracles of God seem to take place within laws of space and time as science knows them.

Physical laws known to science today all start with what are known as "initial conditions." It is impossible to know the state of affairs prior to time = 0, so the physicist lists some fundamental assumptions and some starting conditions he finds reasonable, such as initial temperature, pressure, entropy, etc. He then can run his model and make predictions that are applicable from times greater than $t = 0$. Because of the strange principle known as "time's arrow,"[1] (the observation in science that time always "flows" from past to present to future), it is not possible to do more than speculate about what was going on prior to $t = 0$. The Big-Bang theory (which as we have seen is now quite inadequate to explain things the way they really are) is an example of a theory that assumes that the present-day laws of physics have always prevailed. Thus, many scientists believe they can extrapolate backwards to the beginning from present-day conditions. I, for one, do not accept this assumption.

Some physicists have speculated on the state of the universe for times prior to $t = 0$ in the Big Bang model, but the reader should understand that these are of necessity only speculations, interesting though they may be to examine. But revelation from God given to us in sacred scripture adds additional information to what can be obtained through the five senses, or through intuition, or from theoretical, mathematical analysis. It is not the fault of the present generation that scripture has been neglected by modern science; the roots of this problem go back into the last century and earlier, as is well-documented by Ian T. Taylor (Chapter 5, Note 2).

In creating the universe, God brought order out of chaos by shaping forms and features from previously-created raw materials. For example, He formed man from the dust of the earth and breathed into him the breath of life with the result that man became a living soul. God added energy sources and energy reserves to the created order in stages. Thus, supplies were laid up so the old creation could run on its own "internal" energy reserves for a long period of time.

## God Created the Physical Patterned from the Spiritual

In creation, visible things were patterned after invisible realities. For example, the Tabernacle of Moses and its furniture were copied by Moses after a pattern of the "real" temple in heaven shown to him on Mt. Sinai. God created the spiritual and the physical; however, the spiritual came first. God operated *from* the spiritual world in His acts of direct creation. He provided built-in blueprints in the physical world, incorporated into living things, in the form of incredibly complex genetic codes to allow the self-perpetuation of life. This process of reproduction of living forms would require only energy consumption after the first parent trees, birds, fishes, plants, humans had been completed. The finished creation did not require the introduction of any new outside information to continue replicating and reproducing. As I mentioned, and repeat here for emphasis, exceptions to this rule of no-apparent-outside-interference by God (in the laws of physics)

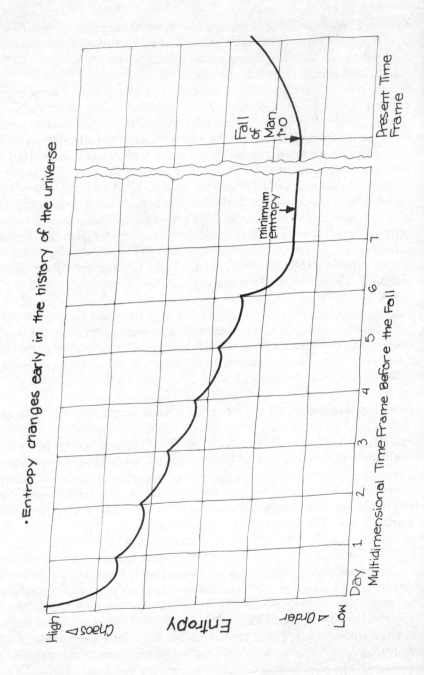

· Entropy changes early in the history of the universe

Entropy

High ◁ Chaos

◁ Order

Low

Day    1    2    3    4    5    6    7    Fall of Man t=0

minimum entropy

Multidimensional Time Frame Before the Fall    Present Time Frame

202

include the subjects of "miracles" discussed earlier and the "New Creation" which is discussed more fully in Chapter 12.

## All the Godhead Involved in Creation

God the Father created the universe by means of a series of thoughts which were then spoken by His Son, as is revealed to us the First Chapter of John's Gospel. The Holy Spirit, as the Third Person of the Godhead, then moved to execute the commands which had called the universe into existence. All three Persons of the Godhead "counselled together" in planning and executing the creation of the universe. On Day One, only one creative word was spoken, **"Let there be light."** Likewise, on Day Two, only one command was given. On Day Three, God spoke two commands which were immediately executed, and on Days Four and Five, one command each. On Day Six, God created the higher forms of animal life as one action, and man, as the second.

Having created man, He instructed man to be fruitful, to multiply and fill the earth, to have dominion over the earth and over all its inhabitants. He planted a garden, towards the East, in Eden (Genesis 2:8), and instructed man concerning special trees, the Tree of Life, and the Tree of the Knowledge of Good and Evil (the fruit of the latter was not to be eaten). Eve was created, I believe, when Adam was created (that is on Day Six) though she may not have been taken out of Adam's side on Day Six, but during a later "week" before the fall. That is, man as created was actually male/female, Adam/Eve. I believe human sexuality originated later with the separation and polarization of what was originally an undivided whole. I assume that God's creative activity was basically finished at the end of the Sixth Day and that He has not tampered with nor "fine-tuned" the universe since then.

I have included (opposing page) a sketch which shows roughly how entropy *may* have changed in the early history of the universe. As can be seen in this sketch each creative act of God lowered the entropy of the universe and produced a discontinuity on the curve. Thus, the entropy of the universe was raised from some initially very high value to a very low value, in steps, until Day Seven. I do not know whether the individual steps were all of the same "size"; probably they were not because of the apparent great amount of creative activity on the Sixth Day especially.

When God rested on the Seventh Day, I assume entropy had its lowest value ever. This means all created energy reservoirs were full, all energy "sinks" were empty, and the universe was in the state of highest order it has ever known. Interaction between the spiritual world and the physical was unimpeded. I assume that time was then flowing in the forward direction towards the future. But the nature of time before the fall was different than it is now, and is beyond our present ability to measure, I believe, until the fall. If the reader will recall that low entropy implies a high degree of order, sophistication, and lots of available "information", as well as much available energy, then the theory that God's creative activity happened in stages will make more sense. This creative activity took six "days" to accomplish, and after that, God ceased from further creative work on the old (original) creation. This is a radically-different view of the origin of all things than modern science presupposes!

## Time as We Know it Began at the Fall

I assume any number of "weeks" of seven days each took place before the temptation and fall of man, and, for the sake of argument, I also suppose that entropy stayed constant during this "epoch". (Note that I am not subscribing to the so-called "day-age" theory of creation nor am I affirming creation in six literal 24-hour days measured by present-day clocks. I also do not believe in the "gap-theory" of an earlier creation, a destruction, and a rebuilding all sandwiched between Genesis 1:1 and 1:2.) I also (boldly!), for discussion purposes, assume there was no radioactive decay in the universe prior to the fall of man and tentatively will also assume that the velocity of light was initially of the order of 10 million times higher than its present value. Setterfield and Norman base their choice of an initial value of c which is $10^7$- (ten million) - times greater than the present value of c on the background x-ray and microwave temperature of space and the red shift data observed today. As mentioned in the last chapter, I think it reasonable to suppose that c-decay was initiated by the fall of Lucifer, the Light-Steward, and if so, c could have begun to decay prior to the fall of man. For the sake of discussion, I will assume c-decay began at the time of the fall of man. This will keep the discussion as simple as possible. Had man *not* fallen, we would today be living in a very different physical universe, as I will attempt to show.

I start my count of time (as we know it now), t = 0, at the fall and exit of man from Eden. Setterfield and Norman's best estimate for this date is 5300-5700 B.C. (gravity clock; that is, dynamical time). This is in close agreement with the dates one obtains from the Hebrew Bible, Septuagint Greek Text, namely 5315 B.C.[2] If we use Alan Montgomery's formula for c-decay, the date of "creation" (or the date c began to decay), would have been 4125 B.C.

My sketch shows hypothetical levels of entropy in the universe during creation week. In this sketch, I have introduced the starting point of history proper (and therefore the epoch of physics as we know it now following t = 0) at a "time" well after the creation of the universe. Earlier I said it was probably difficult to measure time by human standards before man fell, since life on earth was more like life in Paradise than anything you and I have yet known. For this reason I have stylized my sketch to emphasize qualitative, not quantitative, features of entropy changes during creation week.

The fact that pre-history occurred before t = 0 gives our universe an *appearance* of great age, especially if we attempt to study fossils laid down in this time period before radioactive decay commenced. If I am correct in surmising that radioactive clocks began running with the fall of man, then perhaps radioactive decay dates prior to the fall of man correspond to the Archeozoic Era; that is, prior to 600 million years ago in radioactive clock time as Setterfield and Norman have themselves suggested. Before the fall, although "death and decay" as we know it now had not yet entered it is entirely possible that plants grew from sprouts to maturity, and died, giving way to a new generation and that neither man nor animals were immortal in their original bodies.

Considerable amounts of organic matter could have been deposited on the earth and sea floor during this epoch. Adam and Eve had not yet eaten of the Tree of Life; and, hence, their prompt exclusion from the garden after the fall was necessary to prevent them from gaining immortality in fallen, sinful bodies. Therefore, it may have been God's original intention for man to live out 1000 years of natural life, or so, in the original creation and then undergo a staging into a "next life", throwing off his old body and donning a new - like exchanging one suit of clothes for another. This at least is what Bible scholar Ray C. Stedman has suggested to me. Note that we cannot be certain as to whether Psalm 102:26 applies to creation before or after the fall or both:

> "Of old thou didst lay the foundation of the earth,
> and the heavens are the work of thy hands,
> They will perish, but thou dost endure;
> they will all wear out like a garment.
> Thou changest them like a raiment, and they pass away;
> but thou art the same, and thy years have no end.
> The children of they servants shall dwell secure;
> their posterity shall be established before thee."
> (Psalm 102: 25-28).

## How long were Adam and Eve in the Garden?

In the event that a considerable amount of "time" passed in the Garden of Eden, prior to the fall, plants and animals there could easily have gone through a number of generations and fallen into the ground or died in the sea so as to leave a fossil record which we would see as so-called "primitive life" of great antiquity. Did Adam "age" while he was in the garden prior to the fall? I think the answer has to be yes. He grew older and mature but without decrease of strength, disease, or deterioration in body, mind, or spirit. We do not know how "old" Adam was when God breathed the breath of life into his nostrils and man became a living being, but it is safe to assume for purposes of discussion that he was 20 or 30 years "old" and that nature round about him had also an appearance of age since the garden was replete with mature plants and trees and populated, evidently, by adult animals. Seth was born when Adam was 130 years old, (measured by the motion of the earth around the sun), so it is hard to imagine Adam and Eve in the garden for a space of time greater than a few decades. Somewhat later in history, Jesus Christ, a sinless man, was born as a baby, grew up into manhood exactly as we do, and would have lived forever had He not been put to death around the age of 33. Thus, growth from childhood to adulthood to maturity is a feature of the original creation and not something introduced by the fall.

# Earth's Early History: Accelerated Development

Scripture does not tell us how God formed the earth, nor about any additional structuring such as is now present. If we assume the current astronomical model is correct, then the earth formed by accretion of dust and gases drawn together by gravitational attraction and compacted together to form a cool, spinning sphere. Initial gravitational compression would release sufficient energy to raise the core temperature to about 1000 degrees Centigrade.[3] When radioactive decay began, the initial rate of radioactive heating of the earth's interior must have been very high because, as we have seen, the velocity of light was very high and radioactive decay rates are proportional to c. However, in compensation of this high rate of decay, the energy of each emitted photon varies as $1/c$, so the "early heating" would not have been a run-away process as some have supposed.

Thermal conductivity and thermal convection rates vary proportionally to c, allowing more efficient transfer of heat in the interior of the earth. The original earth contained vast amounts of water and dissolved gases as well. I believe these insulated the surface layers, for a time, against too rapid heating. The interior of the earth was then probably rapidly raised above the melting point of all but the outer crust. Core formation probably occurred when the temperature exceeded the melting point of iron. At that point iron and nickel (being both abundant and heavy), sank to the center of the earth, giving up additional gravitational energy in the process. The earth's interior then became differentiated into core, mantle, and crust. Not only was heat available in great amounts, but the high pressures in the mantle rapidly produced metamorphic rocks, granites, etc., at very rapid rates. Viscosity of fluids (such as lava) is inversely proportional to c, so that geological processes in the earth's interior proceeded at a much faster rate in early times compared to now, provided we measure time according to dynamical (gravity) clocks.

Great quantities of gases and water had been trapped in the mantle and crust during the formation of the earth (the Bible says the earth was "formed out of water and by means of water"). A new model for the origin of methane, hydrogen, and petroleum proposed by Professor Thomas Gold of Cornell University a few years ago,[4] fits the known facts about the origin and deposition of natural gas and petroleum better than the old theory of an originally molten earth and petroleum formation solely from fossil sources.

Professor Gold mentions that far more petroleum and natural gas have already been found than can be possibly explained on the basis of the old model of biogenesis. He also notes that existing oil fields lie along fractured zones in the earth's crust where methane contained in the mantle under great pressure can percolate up through existing biomasses to form petroleum and elsewhere, to form carbonate rocks, etc. Gold notes that the major planets of our solar system have large quantities of hydrogen and methane in their present atmospheres so why should earth have been short-changed with regard to these gases that are rare on earth? Hydrogen and helium are the two most common elements in the universe by far; we would expect the earth originally to have had greater quantities of both.

## The Mystery of Earth's Magnetic Field

The earth's magnetic field and its decay is a complete mystery to modern science. Some sort of "dynamo theory" is proposed, but no one has as yet suggested an actual mechanism for the supposedly very ancient field of the earth. The earth's present day magnetic field arises from circulating electrical currents in the core amounting to about a billion amperes and a power level of the order of 800 megawatts. A liquid core is not required to sustain the field once it is established, although a magnetic field once initiated is, of course, expected to decay gradually with time, unless some source for renewal can be found. Theories for the generation of this field by dynamo action in a liquid core have been very unsuccessful in spite of diligent and repeated attempts in recent years to invent some combination of flow patterns that would explain the known magnetic field of the earth. Such theories require a fluid core with complex eddies but no explanation so far proposed fits the basic requirements of freshman physics and the principles of magnetic field generation.

The magnetic field of the earth is also 90% a simple dipole field which gives good reason to choose a very simple model for the origin of the field, not an intricate one, as the more likely to be correct. Without a mechanism of energy coupling from angular momentum to electrical current (so that energy is constantly put back into the magnetic field), the magnetic field would decrease exponentially with time.

In fact, observations of the earth's field for over the past 130 years do show an exponential decrease with a half life (in dynamical time) of only 1400 years. Since it is difficult to imagine an initial field much greater than about 100 gauss (the present level is about 0.5, and 2000 years ago it was about 1.7 gauss), a few brave souls have suggested the magnetic field has only been around 10,000 years or so. I also hold to this viewpoint. The prevailing mind-set of the past hundred years in astronomy and geology is so strong that no one can imagine how the earth's magnetic field could actually be only a few thousand years old. Thus, almost everyone continues to postulate some sort of dynamo driven by fluid motion in the core in a desperate attempt to preserve the notion of the earth's great age and a uniformitarian model for earth's history. Very high magnetic fields are, of course, found in some stars. A straightforward, naive explanation for the earth's magnetic field is that it could have been turned on impulsively during the early core melting and overturning due to rapid radioactive heating shortly after the fall. The earth's field cannot be due to natural magnetism in iron since this effect disappears above the Curie temperature (about 500 degrees C), which occurs less than 30 kilometers into the crust.

The interaction of the earth's magnetic field with the solar wind and particles from solar storms produces the polar auroras on earth and also provides some shielding from (generally harmful) cosmic radiation. There is new evidence that the weather is affected by solar activity coupling with the field. It might be possible that a giant solar flare from the sun soon after the fall of man could have induced the presently-observed magnetic field.

The moon, at present, has been found to have virtually no magnetic field (less than one ten-millionth that of earth). Presumably its initial field has decayed to zero since that body has only 1/64th the volume of the earth. Moon rocks brought back by astronauts show the moon did at one time have a significant magnetic field. The onset of radioactive decay melted the moon all the way to the surface evidently, producing the Maria (seas) which suffered earlier cratering from intensive meteor bombardment before they later froze. Thus the moon we see today probably does not have the same surface it had at the end of creation week by any means.[5]

Mars has a very weak field, equal to the earth's field divided by 300,000. Mercury also has a very weak field, but Jupiter and Saturn have intense fields. A good reason for the very high fields in Jupiter and Saturn could be the high core conductivity (metallic hydrogen) and greater size of these planets. Higher electrical conductivity means a slower rate of exponential decay of the field.

If the magnetic field in all the planets was turned on by an initial burst of energy coming from the sun, then the field has died out only a little on Jupiter and Saturn, is still decaying on earth, and has almost vanished on Mars, etc. Then whether or not the core is fluid has nothing much to do with the field. Rather the decay rate is dependent on the core conductivity, size of planet, etc.

The magnetic field of the sun averages only about 1 gauss, about double the earth's weak field. On the other hand, the sun's field soars to thousands of gauss in sunspots and solar prominences. This is because the sun is an ionized plasma and does not rotate at a uniform rate. As a result, magnetic field lines are dragged along near the equator where the rotational velocity is higher. It is easy to see that rotational energy of the sun can thus be converted into intense localized loops of high magnetic field intensity.

Suppose the sun were to flare up in an unusual way every once in awhile, and in doing so release not a stream of particles but a globe of ball lightning or a closed-loop of circulating current having a very intense field. This loop would then act like the primary of a transformer moving in the vicinity of the secondary turns which are the planetary cores. Were this ever to happen, the earth or other bodies in the vicinity would have an induced field generated in them.

## Core Meltdown

If the earth were originally uniform in its composition prior to internal melting and settling of the iron towards the core, then the length of the day now could have been somewhat shorter than it was in the past. The average density of the earth is 5.52 grams per cubic centimeter. Rocks in the crust have an average density of 2.7, and the iron in the core (under great pressure) has a density of perhaps 10-12. If the earth were originally made of un-differentiated material, the density of the original core after gravitational compaction would have been less than 10 (but greater than 5). Much more mass would have been found in the outer layers of the earth than is now located there. To conserve angular momentum, the rotation of the earth would therefore increase as the core became heavier and the outer layers less dense due to internal radioactive melting. Thus the length of the day may now be somewhat shorter than it was before core melting occurred with the earth. The actual length of the year, measured by the time taken for the earth to revolve around the sun once, would be unchanged by this internal re-distribution of mass within the earth. If this occurred, the number of days in the year would be greater now than in the past. (Some ancient calendars contained 360 instead of the modern 365.25 days per year). A significantly rapid acceleration of the earth's rotational velocity would contribute to stress in the crust, possibly producing earthquakes and tectonic events when core meltdown occurred.

## Life and Population Before the Flood

Genesis 4 and 5 give us the history of the Antediluvian world in a highly-condensed account. From the genealogical list, which is quite complete, the time between Adam and the flood of Noah (which occurred when the latter was 600 years old) can be calculated to be almost exactly 1656 years. During this time period, man was much healthier than he is now; the gene pool, less corrupted by subsequent harmful mutations and other defects; and the environment on earth, much more favorable to good health and long life, as can be seen by the recorded pre-flood longevities. New scientific evidence confirms what Bible scholars have suspected: the earth's ancient atmosphere contained a larger fraction of oxygen than it does at present. This could easily have been due to the higher value of c which caused photosynthesis in plant life to be much more efficient than it is now. Of course a warmer average climate in ancient times would also mean a higher rate of oxygen generation by plants.

At age 65, Enoch became the father of a son whom he named Methuselah, a name which means "when he dies it (the flood) shall come." Enoch went on to walk with God another 300 years and was taken up into heaven by God without dying. Methuselah survived to age 969, the oldest man who ever lived. True to prediction, the flood came the year Methuselah died.

Population growth of human beings can readily be calculated using the formula,

$$P_n = \frac{2\,[C^{n-x+1}]\,[C^x - 1]}{C - 1}.$$

In this formula $P_n$ is the population after n generations beginning with one man and one woman; n is the number of generations (found by dividing the total time period by the number of years per generation); x is the number of generations alive. If people live to see their grandchildren, then c = 3. C is half the number of children in the family. If each family has only two children, the population growth rate is zero, but a more reasonable and conservative number of children per family is 2.5. Allowing for famine, disease, war, and disaster, a few sample calculations will show that the earth's population could have easily reached several billions of people between the time of Adam and the time of the flood. Genesis 4:21-22 gives suggestions of the development of music and advanced technology during this period. Surely human *culture* before the flood was dazzling in comparison to what we now today, even though evil eventually increased to the point of that civilization's self-destruction!

The same formula used above can be used to show the absurdity of evolutionary time scales for mankind. In one million years, if n = 23,256 generations, C = 1.25, and x = 3, the present population of the world would be

$$P = 3.7 \times 10^{2091} \text{ persons}$$

In contrast the total number of electrons in the universe is only $10^{90}$! (These examples are taken from Reference 2, Chapter 5).

The Genesis account suggests that relatively large quantities of water were contained in the upper atmosphere of the earth. That this is reasonable has been shown by calculations and computer modelling.[6] In fact, a close-to-satisfactory mathematical model describing pre-flood conditions in the upper atmosphere of the earth has been developed in recent years. Probably 40 days and 40 nights of rainfall alone would result in less than 150 feet of water on the earth, so most of the water needed for a flood covering the entire earth evidently came from the opening of the so-called "fountains of the deep". These large subterranean water sources were probably entrapped in the earth during its formation. The land masses of earth are now redistributed in size, shape, and average height compared to the depths of the oceans. There is, at present, more than enough water to cover the earth if the land masses were leveled to fill the ocean deeps.

## Early Climate and Weather Patterns

The presence of substantial amounts of water and/or ice in a canopy in the upper atmosphere would raise the surface temperature of the atmosphere all over the earth, due to the greenhouse effect, resulting in temperatures near the poles well above freezing and not too much higher at the equator. The substantial greenhouse effect before the flood would also smooth out what we now know as seasons. Temperate-climate plant and animal fossils are, of course, well-known in Alaska and Siberia, and huge coal deposits and coral reefs are found in Antarctica. They must have come from a period in earth's history when climates were nearly uniform world-wide. The polar regions were free of ice caps. Earth was evidently free of storms, winter, snow, hail, and even rain before the flood. Rather, the ground was watered daily by heavy dew according to Genesis. Calculations show that the atmospheric pressure may have been higher than at present, but the sky would still be transparent and without clouds so that the stars could be seen. We must imagine that the one large land mass that existed before the flood had a much different appearance by far than we now observe in the geography of the earth. However, a recent article purports to have located the area of the Garden in Eden[7] on the basis of satellite imagery, so the flood has probably not totally removed all traces of the old world landforms.

Several excellent accounts of what life on earth might have been like before the flood have been written both by collecting myths and legends[8] and by careful interpretation of Scripture.[9] The earth was not a garden - the Garden to the East in Eden had been planted by God - the rest of the earth was to be subdued by man: cultivated, explored, utilized, and managed. Although the land yielded its crops less readily and brought forth also "thorns and thistles" as a result of the fall, yet the earth must have been bountiful all year around. There were evidently no deserts, no perpetually snow-covered wastes, and probably no very high mountain ranges before the flood.

# The Flood Triggered by "Natural" Events

Setterfield and Norman suggest the flood came at the point in time when radioactive heating of the earth's interior caused a great rupturing of the crust (along the presently identifiable rifts in the earth's crust). The result was a great eruption of the large amount of primeval water out of the crust, sufficient in quantity to flood the earth with some help from the rains coming from above. The collapse of the vapor canopy may have been caused by extraordinary amounts of dust released into the stratosphere by great volcanic explosions as hot magma reached the near-surface waters and was explosively blown high into the sky. If a part of the vapor canopy (over the North and South poles) had been ice, not water, as several researchers have suggested, then the Arctic and Antarctic could have been subjected to sudden deep freezes, entombing thousands of temperate-zone mammoths in Siberia and initiating an ice age. The extreme freezing at the poles would not necessarily be noted in scripture since the text concentrates on the area where the Ark of Noah landed.

The Ark was as large as a good-sized ocean liner, and based on fairly reliable eye-witness reports that it has been sighted during this century, half a dozen expeditions every summer go to Mt. Ararat in Turkey on the Iranian and Russian borders. Should the Ark be located, it might well be found to contain ancient records and artifacts from before the flood which could tell us much about the Antediluvian society.

It is quite possible, Setterfield and Norman suggest, that the extinction of the mammoths in Siberia and the dinosaurs came from another catastrophe after the flood of Noah, in the days of Peleg. I personally tend to favor this view as well.

The New Testament says that the teaching and preaching of the good news of God's grace and mercy, for at least a hundred years before the flood, brought none but eight persons into a personal relationship with the Son of God. We must assume that the message reached the entire ancient world, probably more than once. The flood was surely universal if only eight persons survived out of a race of many millions or even billions of persons living in that ancient civilization.

## Demonic Invasions Before and After the Flood

Genesis, Chapter 6 teaches, I believe, that a great demonic invasion came into the human race prior to the flood because of some sort of sexual intercourse (or demonic possession) involving angels ("the sons of God") and human women. (The Bible does not teach that angels are sexless, only that they do not marry in heaven.) Whatever kind of illicit interaction occurred between a group of fallen angels and the "daughters of men," the offspring were a race of giants, the Nephilim. These legendary beings give rise to many ancient stories of the demigods of Norse, Greek, Roman, and other mythologies.

The angels who were so involved, Jude says, had **"left their proper dwelling places"** and engaged in immoral sexual activity not unlike that found in Sodom and Gomorrah. This group of angels was removed by the Lord from the heavenly places at the time of the flood, and they are now bound in a prison (Tartarus) awaiting the final judgment. Thus God removed one group of angels from Satan's ranks, and from a place of influence among men, at the time of the flood.

However, another demonic invasion found its way into the family of Ham, after the flood, resulting in the extreme sexual depravity of the Canaanites and the rise of a second group of giants, the Anakim. Goliath, slain by David, was one of the latter giants. (I Samuel 17) These very large men frequently were born with extra toes and fingers. Evidently their genes had been damaged and their hormonal balances thrown out of kilter, causing them to grow extraordinarily large as well.[10] Og, King of Bashan, slain by Moses, had an iron bedstead measuring 9.5 feet long by 6 feet wide. (Deuteronomy 3:11) He was one of these post-flood giants.

If Satan "tampered" with the genes and chromosomes of some of the descendants of Adam, then it is not beyond reason that he deranged some members of the animal creation also. What God intended as large lizards could have been the huge animals who became extinct at the end of the Cretaceous period. In any case, men and dinosaurs were contemporary at *some* point in time. Large sea creatures and huge land animals, now extinct, are mentioned at various places in the Bible. These were normal creatures brought into being on the Sixth Day when man was also created. So we cannot say for certain that Satan meddled with the animal creation as he apparently did with man in producing the Anakim and Rephaim.

The fact that many dramatic events were taking place at the time of the Flood of Noah is confirmed in the New Testament by the Lord Jesus and by the Apostle Peter. The Flood was an extensive divine intervention in human affairs. It is quite possible, therefore, that some of the laws of physics became permanently altered at the time of the Flood. If so, there could have been *another* discontinuity in the way the physical world now operates dating from the time of the flood. I have not considered this possibility in the model presented in this book.

The antediluvians lived to ages just short of 1000 years, but after the flood, maximum lifetimes decreased in a smooth curve over several generations until they levelled off at ages more like the present 70 or 80 years of human life expectancy. (Of course during periods of spiritual decline, overpopulation, and pestilence, human longevities have fallen short of the promised 70 to 80 years). Scripture declares that the reduction in human longevity came as a judgment of God, though the mechanism by which this occurred in the cells of our body is not presently very well known. It is often claimed that a vapor canopy over the earth before the flood shielded mankind from the harmful effects of excessive cosmic radiation (which produces genetic damage). It is quite probable that the vapor canopy did provide some healthy protection from dangerous cosmic rays, but probably not enough to account for the observed change in longevities.

## Biological Changes Resulting from c-Decay

Decreasing velocity of light, as a function of time, has consequences for processes going on in living systems which biologists have not yet had time to examine. Since the rate of decrease in c with time was very rapid at the time of the flood, it is possible that metabolic processes were effected by c decay alone so that human lifespans fell quickly after the flood to the present values of 70-80 years. For example, diffusion processes (such as the rate of flow of gases and molecules into and out of cells) are proportional to c, and both heat conduction and metabolic rates are believed to be proportional to c. The nervous system of the human body which is electrical, ionic, and chemical, surely functioned differently in a time when c was higher. It may well be that

216

immune systems which protect us from disease and cancer deteriorated after the Flood. Thus, the velocity of light enters into numerous equations of physics that affect biological processes. Research into the effects of c-decay on living systems opens a whole new frontier of science, in my opinion.

We have no *a priori* reason to assign a smooth, continuous curve to the decay in the velocity of light based solely on evidence from measurements over the past 300 years. Based on Occam's Razor, it is conventional to start with the simplest possible model and add complexities only when necessary in light of additional information. It is possible that the curve has a break in it at the time of the flood. Or perhaps c-decay was uniform during the time of the flood, but discontinuities occurred in some other laws of nature at that time.

It is not my desire to be dogmatic in any of my own assertions, but I would only suggest that a decrease in the velocity of light with time appears to be real, and the initial value of c could have been many orders of magnitude higher than it is now. And if c has indeed changed with time, then changes in other atomic constants, as noted, follow as a consequence, and many processes of physics and nature are affected. Setterfield and Norman have assumed that c-decay has been smooth since creation week. Their curves for c-decay are calculated from the data available for the past 300 years, and they have attempted to determine the maximum initial velocity of light from astronomical data concerning background radiation in space and to estimate the time of "creation" by projecting best-fit curves back to their origins.

Setterfield and Norman have started c-decay during creation week whereas I have assumed that c-decay started with the fall of man. This could have been some significant period of time after creation week as I discussed. Also, Setterfield and Norman have assumed that radioactive decay is a "natural" feature of an unfallen universe; however, I have assumed in my hypothesis that c-decay is a result of and a consequence of the fall of man. I do not see that radioactive decay is necessarily a helpful or valuable phenomenon in the original unfallen creation. It seems to me it is more likely to be a flaw in the creation resulting from the introduction of evil into the spiritual and material realms of the universe.

Part of the work of science is to see how such a set of new assumptions about the world fit the available data better than the

217

old model. I, for one, will feel rewarded if I have stimulated a new look at the world even if a number of my assertions prove later to have been incorrect or incomplete.

## After the Flood

The recovery of the earth after the flood is described in Genesis 9 and 10. Many fine commentaries on the early history of earth before and after the flood have been written, including many hundreds of excellent papers in creationist journals.[11] The first rainbow, given as a sign to Noah, indicates, in the opinion of many fine Bible commentators, that weather and season patterns (as we know them now) had begun. Genesis Chapter 10 describes the repopulation of the entire earth, which was still one land mass, by the offspring of Shem, Ham, and Japheth. Isostatic rebound of the land mass after the enormous weight of the flood waters had drained away, and continuing tectonic processes in the earth driven by heat from the intense (but decreasing) radioactive decay in the crust probably put the finishing touches on many geographical features and some of the mountain ranges we know today. After the flood rapidly-receding waters carved great canyons as large inland lakes drained. Huge sedimentary deposits, thousands of feet thick, were left behind in some areas and possibly *some* of the enormous fossil "graveyards" give evidence of a vast extinction that followed the flood.

Just as the biological implications of c-decay have not yet been addressed by that branch of science, so also geological processes and planetary "evolution" must also be carefully and thoroughly reexamined, in my opinion. The same applies to astronomical, cosmological, molecular, and atomic physics! After that we can begin to understand better the history of the earth in dynamical time, a history which encompasses, apparently, only thousands of years, not millions or billions as has been supposed for the past century and a half by the majority of scientists in the Western world.

It is possible that the laws of physics were also changed at the time of the flood as well as at the time of the fall of man, as I have suggested, or in connection with the fall of Lucifer. If this is so, then we shall have to take another look at c-decay rates, the Second Law of Thermodynamics, and such in this regard. One factor that would suggest another discontinuity in the operation of the physical world at the time of the flood might be found in possible disturbances in nature caused by the removal of a number of fallen angels by God from their positions in the government of the spiritual realm. They were all "chained" and consigned to a prison, allowing an improvement in living conditions for man after the flood. As we have seen, angels play a role in God's government and regulation of the physical universe as well as in God's moral government of society and nations.

## Other Disasters Since the Flood

The flood of Noah was not the last and only disaster earth has known. Setterfield and Norman note that a major wobble in the earth's axis occurred in 2345 B.C. ± 5 years, as thoroughly documented from ancient legends and records by Prof. G. Dodwell, late Government Astronomer for South Australia. Dodwell's monumental study has not yet been published; however, the quality of his research was thorough, and I, for one, feel his findings and conclusions are sound.

The date 2345 B.C. probably corresponds to the time of Peleg, Genesis 11:16-19, I Chronicles 1:19). (The name Peleg means *division* or *watercourse*). Of Peleg it is said in scripture that **"in his days the earth was divided."** Since the name "peleg" gives us our modern words "pelagic" and "archipelago," a number of Bible scholars have suggested that what we now call "continental drift" occurred very rapidly in the days of Peleg.

The lava under the crust of the earth would have had minimum viscosity at this time, according to Setterfield. This is because the radioactive heating of the earth's interior, by 2345 B.C., could have melted the lava just beneath the crust of earth's one continent. At the same time, the viscosity of lava (which is given by an equation containing "c") had a very low value, about equal to that of water.

Setterfield and Norman surmise that the impact of a large meteor or comet striking the earth at that time may have sufficed to split the continents apart and initiate the drifting apart of the land masses to form the continents we know today. Dodwell's data seems clearly to record the recovery of the earth from some sort of impact by an extraterrestrial body. The continents still appear to be still drifting apart a few centimeters a year at the present time, but of course this could be just the tail-end deceleration of a once very rapid motion. The "Dodwell" event, (that is, an asteroid or comet colliding with the earth) could be the same event suggested as the reason for extinction of the dinosaurs by geologist Walter Alvarez and his father physicist Luis Alvarez of the University of California at Berkeley.

According to Setterfield and Norman, the extinction of the dinosaurs seems to coincide with the world-wide disaster studied by Dodwell as a wobble in the tilt of the earth's axis noted in ancient records. The radioactive clock date corresponding to true calendar time of 2345 B.C. is estimated to be 71.6 millions of years before the present. The atomic date of the flood, 256 million years ago, in the Paleozic era, corresponds to a gravity-clock date of 3000 B.C. in the Setterfield-Norman model. Thus the splitting apart of the continents occurred about 650 years after the flood.

Although the division of the continents could have been "moderately" non-destructive, due to the slippery lava under the land mass, considerable loss of animal and human life probably followed. Excessive volcanic activity following the new fissuring of the earth could have released much dust into the earth's upper atmosphere producing severe climate changes for a few years. As the continents split part, Indians were trapped in what is now North and South America; many others of the Hamitic peoples remained relatively isolated in Africa; the Japethites ended up in Europe, etc. Certain species died out and were lost everywhere except in Australia, and so forth.

Perhaps ancient earth legends can be pieced together about this time of continental break-up in the same way that many scholars have tried to trace pagan mythologies back to the Flood of Noah. This model assumes that continental break-up and drift occupied a period of time of only a few years. At first the motion was rapid, but as the underlying lava cooled and became less viscous, continental drift decreased to nearly zero. The splitting of the continents in Peleg's time would have isolated men and their characteristic, differing cultures from one another. Peleg lived about the time of the Tower of Babel when earth's one language was sundered into many different tongues - an act of God to further divide men from one another for their own good.

Probably major climate pattern changes resulted from the new land and sea distributions, and today's weather patterns were gradually set up since these depend on the geography of the land masses and oceanic bodies as well as upon solar insolation and ocean currents. As everyone knows, the vast amount of water now contained in earth's ice-caps and glaciers has lowered the ocean levels considerably since the days of Peleg. It is not unreasonable to suppose that volcanic activity and other outgassing processes changed the constituents of the earth's atmosphere permanently at the time the earth was divided.

## When the Mediterranean Was A Dry Salt Bed

A recent, very credible scientific book[12] documents the fact that the Mediterranean Sea was once an inland sea that dried up completely and only partially refilled in cycles before drying up again. Large deposits of salt were found everywhere under the bottom of the present sea during a recent, extensive core-drilling program. The evidence also points to a dramatic eventual refilling of the Mediterranean by means of a gigantic waterfall at the Straits of Gibraltar.

This refilling began about 5.5 million years ago in radioactive time. Marine deposits under the Nile Valley as far south as Aswan show that the previous salt sea once filled a deep gorge (which is now the filled-in Nile Valley). As the Mediterranean Sea refilled, men from the Fertile Crescent area probably moved into the lands around the Mediterranean Sea to settle there, and of course the climate there began to be hospitable as we know it today. The filling of the Nile with fresh water sediments about the same time made Egypt inhabitable, so we can in principle begin to estimate the time the ancient Egyptians settled along the Nile. These events seem to have taken place not long after the division of the continents in the days of Peleg. Ancient Egyptian civilization seems to have appeared suddenly out of nowhere about this time in early history.

A serious look at the fossil records indicates that there appear to have been other great die-offs in animals and plants in history *since* the flood and the days of Peleg. Thus, other major destructive events seem to have occurred since then. Until recently many creationists have tentatively assumed that the fossil record was the record of the flood only. This has been incredulous to the average geologist or palentologist.

I believe it must now be said that more than one catastrophe or disaster in earth's history has left its marks in the geological column and the fossil record. Mapping these geologic events and locating them in dynamical time presents a wonderful challenge for researchers in our time. The fact that a number of disasters of all varities, large and small, have been known in recorded history has not gone undocumented.[13]

## The Causes of the Flood and other Disasters

It is important to note that had man not fallen, there would have been no flood. The flood was a judgment of God on an unrepentant race who were almost all unwilling to be restored to fellowship with God by the time of Noah. God was the cause of the flood, but if He used radioactive decay as the physical process that brought about the flood and other subsequent disasters as consequences of human sin, then our assumption that radioactive decay was turned-on, or triggered by the fall would seem to be a reasonable one.

The apparent rate of c-decay was very rapid at first, according to any of the equations given in Chapter Six or the curve presented there. During the time interval when c was falling at the greatest rate, the law of conservation of energy requires that the universe as a whole "spend" the excess energy available because of the rapid collapse of space. This could be done by allowing the previous state of high order and complex organization in the universe to fall rapidly in the direction of chaos.

This is the same thing as saying that the entropy of the universe increased rapidly after the fall of man. I have already suggested that "power flow" from the spiritual world into the physical world decreased as a result of the fall of man. It also seems reasonable that the "regulatory" activity of the angels, (which would ordinarily preserve harmony and order in God's universe) became less effective, especially with one-third of the angels breaking into full-scale cosmic revolt against the Lord. This may imply that astronomical events, such as supernovas, quasars, and black holes came into existence when c began to drop, when the universe was forced to "get rid of" excess energy. Even in the absence of matter, empty space possesses potential energy (measured by the so called "Stress-Energy Tensor"[14]). As God "stretched out" the heavens, empty space was invested with energy which it later began to lose when the fabric of space "relaxed." The release of energy stored in space could be expected to produce vortices and instabilities in space. These instabilities would not have occurred if the expanse of space had remained at a constant limit after creation. (As a side note, it is interesting that most of the total energy in our universe exists in the form of rest mass. The total kinetic energy of the entire known universe is less than the energy-equivalent of the moon's mass.)

It seems to me reasonable to suppose that many of the perplexing mysteries of the heavens seen today by astronomers may be part of enormous disruptions in the original energy distribution of the universe. Many of these events may have happened since creation week, and give evidence of havoc, wasting, and ruin brought about on a "cosmic" scale by the active evil present in the universe at all levels. I have in mind black holes, supernovas, superluminal sources,[15] quasars (some of which appear to be travelling at 3 to 20 times the velocity of light), and the like.

�֎

## Notes to Chapter Ten

1. See for example, Richard Morris, *Time's Arrows* (Simon and Schuster; New York, 1980), which is an excellent, recent book on this subject. Morris includes a very good bibliography in his book.

2. Setterfield and Norman discuss the differences between the Septuagint (LXX) and Masoretic dates for the time of Adam, and indicate their reasons for preferring the LXX in the Supplement to their research report, Reference 2, Chapter 6.

3. Frank Press and Raymond Siever, *Earth* (W. H. Freeman and Company; San Francisco, 1974), p. 517ff.

4. See "The Origin of Petroleum" by David Osborne, *The Atlantic Monthly*, (February 1986); "Models of the Earth's Core" by D. J. Stevenson, *Science* Vol. 214, 6 (November 1981), pp. 611-619; "The Deep Earth" by Thomas Gold, *Geophysics: The Leading Edge of Exploration* (July 1984).

An interesting recent book that shows remarkable evidence about the early history of the earth based on laboratory studies of diamonds is *Ancient Diamond Time Capsules, Secrets of Life and the World* by Charles E. Melton, a Professor of Physical Chemistry at the University of Georgia (Melton-Giardini Books Company; Rt. 2, Box 18; Hull, Georgia 30646).

5. See John C. Whitcomb and Donald B. DeYoung, *The Moon: Its Creation, Form and Significance* (Baker Book House; Grand Rapids, 1978).

6. A definitive early study is that of Joseph C. Dillow, *The Waters Above* (Moody Press; Chicago, 1982). Refinements have since been offered by Gary L. Johnson, "Global Heat Balance With A Liquid Water and Ice Canopy" *Creation Research Society Quarterly*, Vol. 23 (September, 1986).

7. Dora Jane Hamblin, "Has the Garden of Eden Been Located at Last?" *Smithsonian*, Vol. 18, No. 2, (May, 1987).

8. Marcus Von Wellnitz has compiled a number of fascinating legends and traditional accounts of the flood in an article entitled, "Noah and The Flood: The Apocryphal Traditions" *Creation Research Society Journal*, Vol. 16, (June, 1979).

9. John C. Whitcomb's book, *The Early Earth* (Baker Book House; Grand Rapids, 1986) is a recent good book on the subject. The author gives an excellent bibliography of other recent books related to this subject. See also his book *The World That Perished* on the flood itself, (Baker Book House; Grand Rapids, 1986). Although written many years ago, John C. Whitcomb and Henry M. Morris' *The Genesis Flood* (Presbyterian and Reformed Publishing Co.; 1961), remains a classic in this field.

10. Scientific aspects of the book of Genesis are emphasized in Henry M. Morris' excellent commentary on Genesis, *The Genesis Record*, (Baker Book House; Grand Rapids, 1976). I highly recommend the *Proceedings of the International Conference on Creationism* held in Pittsburgh in 1986. The two-volume set is available for $35.00 from I.C.C.; P.O. Box 17578; Pittsburgh, PA 15235.

11. See W.A. Berggren and John A. Van Couvering, Editors, *Catastrophes and Earth History* (Princeton University Press; 1984).

12. Kenneth J. Hsü, *The Mediterranean was a Desert: A Voyage of the Glomar Challenger* (Princeton University Press, 1983).

13. Steven A. Austin, *Catastrophes in Earth History: A Source Book of Geologic Evidence, Speculation and Theory* (Institute of Creation Research; El Cajon, CA., 1984).

14. See Charles W. Misner, Kip S. Thorne, and John Archibald Wheeler, *Gravitation* (W. H. Freeman and Company; San Francisco, 1973), Chapter 5.

15. Gerritt L. Verschuur, *The Invisible Universe Revealed: The Story of Radio Astronomy* (Spring-Verlag; New York; 1987) and J. Anton Zensus and Timothy J. Pearson, Editors, *Superluminal Radio Sources* (Cambridge University Press; 1987).

# Chapter Eleven

## The Undoing of Cosmic Evil

I hope that I have convinced the reader from my brief treatment of Biblical themes that both the heavens and the earth are now filled with an active and pernicious evil and that both realms of creation have become flawed as a result. I suggest that some of the laws of physics we now take for granted were different in the past. Evil in the heavens means that malevolent spiritual beings, having great influence in the universe serving Satan, have access to the throne of God and to territories beyond the earth.

As the "god of this world", Satan now rules in the activities of men; however, only with permission from God, and also in subjection to God. This rule of evil in human affairs is temporary and coming to an end. In fact, the downfall of the Evil One has already been accomplished in eternity. We who live constrained in time can rest assured that a bright, new world lies ahead for all who follow Jesus as Lord. A just and holy God cannot tolerate the present world situation forever, but must , and will, intervene and change the status quo. One such intervention has already occurred, at the time of the Flood of Noah. God's next moves will be more grand, terrible, and awesome indeed.

### Man's Three Enemies

On earth, man is fallen, and every one of us suffers from the wasting ravages of sin in his or her own person. Even those who know Jesus Christ and have been spiritually regenerated live in physical bodies that are not yet redeemed. Non-Christians are, in fact, said to be "dead in sin," so in one sense they cannot be expected to live moral and godly lives by nature.

Becoming a Christian does not eliminate sin, rather, conversion to Christ is the time the real battle begins! Christians find themselves subjected to temptations and inclinations towards evil through three mechanisms: what the Bible calls, "the flesh, the world, and the devil." Neither the body, nor matter, nor things in the material world are, in and of themselves, evil.

The biblical term "the flesh" might better be translated "the self-life." The seat of the flesh lies in the as-yet-unredeemed physical body of man, but it is Satan who energizes and empowers these lusts of the flesh seeking to draw us away from dependence upon the indwelling Lord Jesus. The flesh has both "good" and "bad" aspects and essentially always springs to life when a Christian tries to live his new life by self-effort rather than by dependence upon his indwelling Lord. Christians are free to "walk after the flesh" but admonished instead to "put to death the deeds of the body" and to "sow by the Spirit" since there are inevitable consequences for evil or for good depending on all our daily choices.

The "world", as the New Testament uses the term, is not the world of nature, but culture, custom, tradition, and society as dominated by Satan. A thousand allurements from the world assail us daily in the form of philosophies, attractions to pleasure and to self-indulgence, diversions to materialistic pursuits, and distractions intended to cause us to walk by sight rather than by faith.

The good news, the "gospel" truth, is that God has already solved the problem of evil in both dimensions, that is, "in heaven" and "on earth". This is why the theme of victory, triumph, and hope pervades the New Testament. **"...I would have you wise as to what is good and guileless as to what is evil; then the God of peace will soon crush Satan under your feet."** writes Paul (Romans 16: 19-20) To the Corinthians he says, **'For he (Jesus) must reign until he has put all his enemies under his feet. The last enemy to be destroyed is death. 'For God has put all things under his feet.'''** (I Corinthians 15:25-27)

## Why God Became a Man

The solution to the cosmic problem could not, cannot, and did not come from human skills, ingenuity, or meritorious efforts. It is the Creator Himself who made plans from the beginning of time for the solution of man's terrible plight. In due season, **"in the fulness of time"**, God executed those plans according to His own timing and pre-planning. What God did was to enter the human race as a perfect, sinless man and to become a substitute for each

228

one of us, a sin-bearer, a reconciler, **"the Author and Finisher of our Faith."** One of the great passages of Scripture which describes the "kenosis" or "self-emptying" of the Son of God is found in Paul's letter to Philippians:

**"Let this mind be in yourselves which was also in Christ Jesus, who though He was subsisting in the essential form** (morphe) **of God, yet thought it not a prize or treasure to be grasped at, to be equal to God, but emptied** (ekenosen) **Himself** (voluntarily) **taking the form of a servant, having become in the likeness** (homoiomati) **of men, and being found in fashion** (schemati) **as a man, He humbled Himself** (adopted a low profile or took a lowly position), **becoming obedient until death, even the death of a cross. Therefore also God highly exalted him and gave to Him a name that is above every name, that at the name of Jesus every knee should bow, in heaven, and on the earth, and under the earth, and every tongue should confess that Jesus Christ is Lord, to the glory of God the Father."** (Philippians 2:5-11 My paraphrase)

This passage is of such profound importance to understanding the nature of God and the incarnation of Christ that great debates raged in the early church over the issue of whether Christ had one nature or two (the problem of the "hypostatic union"). Nearly all Christians today agree that this passage states that Jesus Christ is fully God and fully man. From His own words and life as seen in the Gospels, we can see that Jesus lived His entire life on earth by faith in total dependence upon the Father who dwelt in Him, and that He did not, while He was on earth, exercise His sovereign power as God the Son. This right and privilege (that of acting as Sovereign God), in addition to His exalted and splendrous place beside the Father, were temporarily and voluntarily set aside. Thus, the Son of God might become a man and die as a substitute for the sins of the world. Having now accomplished the work of the cross and having been raised from the dead, Jesus now sits at the "Right hand of the Majesty on High". He has actually gained a more exalted position than He held before, if such a thing is possible for us to imagine.

## The Seat of Original Sin

In his book, *The Seed of Woman,*[1] Arthur Custance makes a case that original sin may very well be transmitted biologically from generation to generation through the male sperm, rather than through the female ovum. Custance takes great pains to defend his premises thoroughly, and his careful scholarship in recent decades earned him a respected place in conservative Christian circles before his death a few years ago. The virgin birth allowed Jesus to be born of Mary free from all sin so as to become a "lamb without spot or blemish," "tempted in every way, just as we are, yet without sin." Scripture also speaks of Jesus as "The Lamb slain before the foundation of the world."

Thus the cross of Christ is the intrusion into our limited space-time domain of an event, or happening, a transaction between the Father and the Son, which really takes place in eternity, outside of time. Custance's argument is a very reasonable one and helps us to understand why the blood line of the promise through legitimate heirs from Eve down through Mary is uninterrupted, while only the Kingly promise is preserved from Abraham to Joseph. The conception of Jesus in the womb of Mary by the Holy Spirit interrupted the chain of genetic links beginning with the fall, allowing a descendant of Adam to be born into the world free from original sin. The perfect obedience of Jesus during His life on earth *also* was necessary to assure that He reached the cross as a fully qualified sin-offering. Scripture emphasizes the humanity of the Messiah as fully as it does His Deity. The Old Testament is replete with references to the Messiah as the "root out of dry ground", "the seed of David", "the suffering servant of the LORD",[2] and so on. (For the reader who may be unfamiliar with the claims of Scripture about itself and the many Biblical statements about Jesus, the Messiah, I have included some convenient references in the Appendix.)

## The Accomplishments of Jesus

In Scripture, when God has something important to say, it is sometimes repeated. Usually one repetition is sufficient to tell us to pay attention, as when Jesus would begin a statement with the words "Truly, Truly I say to you..." As contemporary theologian R.C. Sproul has noted,[3] occasionally Scripture repeats something three times in a row to make certain we understand, for example, the central importance of the holiness of God among all His other attributes. Only a few times does Scripture repeat something three times for emphasis. When it comes to the life of Jesus - His temptations, betrayal, trial, death and resurrection - four, not three, Gospels were written. Surely this strategy by the Holy Spirit is intended to help us see the great importance of God becoming a man, to see that "...in Christ God was reconciling the world to himself, not counting their trespasses against them..." (II Corinthians 5:19)

John R. W. Stott has recently published a thorough, careful, and most helpful book[4] on the cross of Christ which I recommend highly. The cross of Christ is all-too-frequently neglected or even crowded out of Christianity by other less "offensive" aspects of theology and Bible study, but this can be done only at the cost of losing the most important core message of the Christian faith. Not only is the subject of the Cross all about the death of Jesus on our behalf, but also it points to the fact that we, too, must be put to death, in Christ, on that same cross, to gain eternal life. Our crucifixion *with* Christ calls attention to the fact that there is nothing in the old creation, in the first Adam, that can be saved apart from death. Paul writes, **"I have been crucified with Christ; nevertheless I live, yet not I, but Christ lives in me, and the life I now live in the flesh I live by the faith of the Son of God who loved me and gave himself for me. I do not nullify the grace of God, for if justification came by the law, then Christ died to no avail."** (Galatians 2:20-21 My paraphrase)

While writing to encourage the Christians in the early church at Colossae, the Apostle Paul reveals to them some of the mighty once-for-all-time accomplishments of Jesus on the cross:

**"As therefore you received Christ Jesus the Lord, so live in him, rooted and built up in him and established in the faith, just as you were taught, abounding in thanksgiving. See to it that no one makes a prey of you by philosophy and empty deceit, according to human tradition, according to the elemental spirits (*stoicheia*) of the universe, and not according to Christ. For in him the whole fulness of deity dwells bodily (permanently), and you have come to fulness of life in him, who is the head of all rule and authority. In him also you were circumcised with a circumcision made without hands, by putting off the body of flesh in the circumcision of Christ; and you were buried with him in (the) baptism, in which you were also raised with him through faith in the working of God, who raised him from the dead. And you, who were dead in trespasses and the uncircumcision of your flesh, God made alive together with him, having forgiven us all our trespasses, having cancelled the (legal) bond which stood against us with its legal**

demands; this he set aside (blotted out), **nailing it to the cross. He disarmed** (stripped of power and authority) **the principalities and powers** (in the heavenly places) **and made a public example** (spectacle) **of them,** (bodily) **triumphing over them in him."** (Colossians 2: 6-15)

## The Cosmic Struggle on the Cross

This passage reveals that not only did Jesus take upon Himself the sins of the world when He died for us on the cross, but He also met fully the onslaught of demons, fallen angels, and all the power of evil forces in the heavens as well, disarming all of them completely. His victory over man's greatest enemy, death, is boldly stated in the letter to the Hebrews: **"Since therefore the children share in flesh and blood, he himself likewise partook of the same nature, that through death he might destroy him who has the power of death, that is, the devil, and deliver all those who through fear of death were subject to lifelong bondage."** (Hebrews 2:14,15) In speaking to the Apostle John from the heavens, Jesus sent these words to mankind: **"Fear not, I am the first and the last, and the living one; I died, and behold I am alive for evermore, and I have the keys of Death and Hades."** (Revelation 1:17-18)

Jesus, *on the cross*, also won back any and all claims Satan had on man, or the earth, or as an authority of any kind in the heavens. If, for example, Satan claimed to hold the title deed of the earth (having gained it because of Adam's fall) that deed now belongs to Jesus as one of the results of His work on the cross. (This is known as the "ransom" work of Christ on the cross). Satan's destruction, too, was accomplished on the cross, outside of time, for which we await now only the final outworkings in history. This unseen and invisible victory over cosmic evil on the cross is yet another reason why Jesus alone is qualified to receive from the Father all honor and power and glory:

**"And I** (John) **saw in the right hand of him who was seated on the throne a scroll written within and on the back, sealed with seven seals; and I saw a strong angel proclaiming with a loud voice, 'Who is worthy to open the scroll and break its seals?' And no one in heaven or on earth or under the earth was able to open the scroll or to look into it, and I wept much that no one was found worthy to open the scroll or to look into it. Then one of the**

(twenty-four) **elders said to me, 'Weep not; lo, the Lion of the tribe of Judah, the Root of David, has conquered** (overcome), **so that he can open the scroll and its seven seals.'** And between the throne **and the four living creatures and among the elders, I saw a Lamb standing, as though it had been slain, with seven horns and with seven eyes, which are the seven spirits of God sent out into all the earth; and he went and took the scroll from the right hand of him who was seated on the throne."** (Revelation 5:1-7)

## Jesus: Great High Priest and Perfect Sacrifice

I want to consider, briefly only two aspects of the death of Christ that show something of the mystery of His death and the suffering He took onto Himself for our sake. As noted, I am drawing from the studies of Arthur Custance. The death of Jesus on the cross took but six hours as measured in dynamical time. Jesus was, for the first three hours on the cross, our Great High Priest. From noon till 3 P.M., during which time a strange and terrible darkness came over the earth, the Priest became the Sacrifice. I have diagrammed this event on the following page, as if it were seen by an observer in Jerusalem at the time. I have included in this sketch the seven last words Jesus spoke from the cross for reference purposes.

The reader should immediately reflect upon our discussion earlier about the nature of time and eternity and realize that what was (for us) three hours' suffering by Jesus in total estrangement from the Father was for him *an event in eternity* which never ends.

## How and When Did Christ Descend Into Hell?

It is because the sufferings of Christ extend out of time into eternity that many have come to believe that Christ descended into hell. For example, one of the statements of faith, a version of the Apostles' Creed which one usually recites in a Sunday Episcopalian worship service, has a reference to Jesus Christ descending into hell. At first we might suppose that this refers to a visit by Jesus to the "underworld" which took place after His death on Good Friday, but prior to His resurrection on Easter Sunday morning.

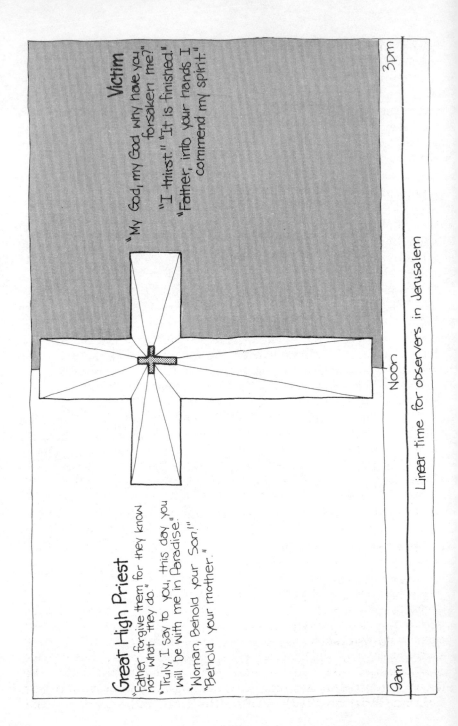

Great High Priest

"Father, forgive them, for they know not what they do."

"Truly, I say to you, this day you will be with me in Paradise."

"Woman, Behold your Son!"

"Behold your mother."

Victim

"My God, my God, why have you forsaken me?"

"I thirst." "It is finished."

"Father, into your hands I commend my spirit."

9am     Noon     3pm

Linear time for observers in Jerusalem

The Apostles' Creed reads:

"I believe in God, the Father almighty,
maker of heaven and earth;
and in Jesus Christ his only Son our Lord;
who was conceived by the Holy Ghost,
born of the Virgin Mary,
suffered under Pontius Pilate,
was crucified, dead and buried.
He descended into hell.
The third day he rose again from the dead.
He ascended into heaven,
and sitteth on the right hand of God the Father almighty.
From thence he shall come to judge the quick and the dead.
I believe in the Holy Ghost,
the holy catholic Church,
the communion of saints,
the forgiveness of sins,
the resurrection of the body,
and the life everlasting. Amen."

This particular creed, known since the Third Century in the Western Church, draws reliably from the New Testament in general. However the phrase "He descended into hell" was evidently taken from an unusually-worded portion of Peter's First Epistle:

"He (Jesus) was put to death in the flesh, but he was raised to life in the Spirit, in which also he went and preached to the disobedient spirits who were in prison in the days of Noah when God waited patiently while the ark was being built...For this is why the gospel was preached even to the dead so that, although they have already been judged in the flesh like men, they might have life in the Spirit like God." (1 Peter 3:18-20; 4:6 My paraphrase)

The first thing to note in this passage is that the Greek word "Hades" - translated "hell" in the Apostles' Creed - is the underworld of Greek mythology, not necessarily the place of permanent punishment of those utterly lost forever. Whatever preaching took place was among the Antediluvians; that is, men and women who lived before the flood of Noah. Some Bible commentaries take this to mean that those who died before the flood were present as spirits in a holding-tank known as Hades and that Christ went to them after His death "in the Spirit" to offer them a second chance. Others have argued that the "spirits in prison" were fallen angels (mentioned by Jude) who were removed from the world scene at the time of the flood and "...**have been kept by him in eternal chains in the nether gloom until the judgment of the great day.**" (Jude 6) Still others argue that Christ preached only to the spirits of the righteous dead who lived before His time, and when He did so, He emptied Hades, leading those who were waiting there out and into Paradise.

## The Gospel Sent Forth in Every Age!

All of these views force the Scripture beyond what it really says, I personally believe. The clearest way to rephrase this passage so that it makes sense to me and takes into account the nature of time and eternity, is this: "In the days of Noah, while the Ark was being built, the Gospel was preached by Christ, who spoke by means of the Holy Spirit through Noah to the men and women of his generation." For God to endure patiently the wickedness of perhaps several billion people while giving them every chance to be saved demonstrates His longsuffering love for sinners and great desire that none should be lost.

In actual fact, only Noah, his three sons, and their wives, (eight persons in all), took heed to the message. The spirits in prison could then be those who had not been regenerated - this is the general condition of mankind after the fall, **"dead through (the) trespasses and sins...following the course of this world, following the prince of the power of the air** (Satan), **the spirit that is now at work in the sons of disobedience."** (Ephesians Chapter 2:1,2) Other scholars comment that the spirits in prison were not fallen men, but angels imprisoned at the time of the flood because of their special wickedness in intermarrying with women so as to produce the antediluvian giants. Noah and his family who responded to the Gospel were saved; all others who heard the good news totally rejected it. There is no Biblical evidence, however, that God offers salvation to fallen angels; and Hebrews 9:27 says clearly that men are not given a "second chance" after physical death, **"...it is appointed for men to die once, and after that comes the judgment."**

In Ephesians 4:9, following a description of Christ's triumphant ascension into heaven there is a parenthetical reference, **"[In saying, 'he ascended,' what does it mean but that he had also**

**descended into the lower parts of the earth.**]' Although the passage makes the descent of Christ to districts of earth a very great descent in comparison with the grandeur of His position in heaven before his incarnation (in line with Philippians 2:6-11), some Bible commentators have taken this passage to be a second New Testament reference to Christ's descent into hell. If this is so, hell is assumed (in the Greek world view) to be a subterranean region or cavern beneath the earth. The Old Testament, however, was vague about what happened after death; *Sheol* meant "death" or "the grave" but little else is specified or revealed.

Only as the New Testament unfolds, do we receive an understanding of "the Lake of Fire" as the place **"prepared for the devil and his angels"** where the wicked (and the devil) are ultimately cast. Likewise we are given to see that the "heavenly places" are the realm of the Spirit into which all believers have already been placed, though our bodies are not yet redeemed. Actually, the New Testament does not speak of waiting rooms for the spirits of either the righteous or the unrighteous dead, nor of an intermediate state between death and resurrection, nor of the spirits of the dead being without a body for more than an instant of time! Although the righteous who have died are said to have **"fallen asleep in Christ,"** this is merely an accommodation to our earthly reference frame to remind us that the terrors and fears of death are gone for us who are in Christ Jesus. Of course, the notion of purgatory is foreign to Scripture, but believers should not take lightly the possibility of great and painful losses at the judgment seat of Christ. And no one should presume that any of us shall enter heaven without a thorough evaluation and full disclosure of the actual quality and content of our lives since we first believed.

When Christ **"dismissed his spirit"**, commending Himself into the hands of the Father, the work of the cross having been finished, the spirit and soul of Jesus as well as that of His companion, the forgiven thief, went immediately into heaven. In so doing, they left our zone of linear space-time and stepped into eternity. Furthermore, Paul says in II Corinthians 5:1-5 that resurrection bodies are already waiting and prepared for all believers in heaven and that to be absent from the body is to be at home with the Lord. Our problem is one of not understanding that we are constrained by our present fallen, mortal bodies to linear time (which always flows from past to present to future) whereas spirits live in the eternal dimension already. Thus, to die physically today and appear at the resurrection of the dead (which may occur 100 years from now in calendar time) means that the intervening 100 years, to the person who is dying, is but **"a moment, the twinkling of an eye."**

Peter tells us in his sermon on the Day of Pentecost, quoting from Psalm 16, that the body of Jesus did not begin to decay in the tomb between the time of His death and His resurrection, although the April weather in Jerusalem at Passover would have been warm and no embalming of the corpse had been done:

**"Men of Israel, hear these words: Jesus of Nazareth, a man attested to you by God with mighty works and wonders and signs which God did through him in your midst, as you yourselves know - this Jesus, delivered up according to the definite plan and foreknowledge of God, you crucified and killed by the hands of lawless men. But God raised him up, having loosed the pangs of death, because it was not possible for him to be held by it. For David says concerning him, 'I saw the Lord always before me, for he is at my right hand that I may not be shaken; therefore my heart was glad, and my tongue rejoiced; moreover my flesh will dwell in hope. For thou wilt not abandon my soul to Hades, nor let thy Holy One see corruption. Thou hast made known to me the ways of life; thou wilt make me full of gladness with thy presence'."** (Acts 2:22-28)

## We All Get to Heaven at the Same "Time"!

In the letter to the Hebrews, after reciting a great list of the works of faith by many righteous men and women in the Old Testament, the writer concludes, **"And all these, though well attested by their faith, did not receive what was promised** (New Jerusalem), **since God had foreseen something better for us** (that is, believers in the present day under the New Covenant), **that apart from us they** (these Old Testament saints) **should not be made perfect."** (Hebrews 11: 39)

If we think of death as the leaving of time and entering into eternity, this passage speaks of the gathering of *all* believers together at some future event. In the experience of any individual who dies, whether today or 4000 years ago, the time interval between death and resurrection is but a moment in eternity. Understood in this light, all believers reach heaven at the same moment, and the Second Coming of Christ coincides with the moment of one's death. That great event in eternity will one day intersect a point in earth-time described for us in I Thessalonians 4:13-18: "But we would not have you ignorant, brethren, concerning those who are asleep, that you may not grieve as others do who have no hope. For since we believe that Jesus died and rose again, even so, through Jesus, God will bring with him those who have fallen asleep. For this we declare to you by the word of the Lord, that we who are alive, who are left until the coming (parousia) of the Lord, shall not precede those who have fallen asleep. For the Lord himself will descend from heaven with a cry of command, (to raise the dead), with the archangel's call (to call Israel back to Himself), and with the sound of the trumpet of God (to transform those believers who are alive at the time). And the dead in Christ will rise first; then we who are alive, who are left, shall be caught up together with them in the clouds to meet the Lord in the air; and so we shall always be with the Lord. Therefore comfort one another with these words."

There is, therefore, in my opinion, no reason to believe that Christ "descended into hell" or Hades *after* He died on Good Friday, and no reason to believe there are "waiting rooms" for disembodied spirits who are waiting for the resurrection. I do not believe there is an intermediate state after death; and in fact I believe no one is in heaven as yet (except our Lord Jesus). So, of course, prayers to Mary or to other saints who have gone before are doubly without meaning.

As Hebrews 12 says, all believers, whether Old Testament or New, are already in New Jerusalem, in spirit, as a great assembled company. We are merely awaiting the revelation, the unveiling of our Lord, and the redemption of our bodies so that we will have eyes to see the invisible world of heaven around us and new bodies that can experience the multiple dimensions of time and space in that realm where God dwells. There Christ sits at the right hand of the Majesty on High. If the reader understands that the "heavenly places" are all around us, and not far beyond the reaches of space, then death is merely the lifting of a veil that separates the physical from the spiritual. All those who are *"in Christ"* are seated with Him, now, in the throne room of God, in His very presence and do not need a spaceship to reach their heavenly home!

# The Eternal Sufferings of God in Christ

We still must consider the full extent of what happened when Christ met Death, *while He was on the cross*, and how He defeated evil forever. So far the discussion has centered on whether or not Christ descended into hell, following his death at 3 P.M. on Good Friday and before His resurrection about dawn on Easter Sunday morning. I take the statement of Jesus, **"Truly, I say to you, today you will be with me in Paradise."** (Luke 23:43) to mean that when He died, Jesus left our time frame and entered eternity; likewise, the spirit of the dying, redeemed thief also left time and entered eternity when he died on Good Friday.

The next event *in eternity* for the spirit of Jesus was His return to reenter His body in the tomb. By means of the mighty power of the Holy Spirit, He then experienced the complete transformation of His body and His resurrection "out from among the dead." In the time frame of earth, these events are separated by perhaps 40 hours, but in eternity they are an immediate sequence of events, one following another. The dying thief was not raised from the dead at the same earth time as Jesus was raised from the dead. However, in his own (the thief's) consciousness, he stepped out of time to join the general resurrection of all the righteous dead which coincides in history with the Second Coming of Christ.

In this sense, I claimed above that neither heaven nor hell are yet populated, and this is an added argument for not praying to the saints or to Mary, for all believers reach heaven at the same "time." The dying thief, Stephen the first martyr, the Apostle John, and all the rest of us will arrive in heaven at precisely the same "instant", experiencing neither soul sleep nor loss of consciousness nor time delay, whether the interval between our death and the Second Coming is a hundred years or one hour. The thief on the cross, in his own consciousness, will *experience* arriving in Paradise the very same day he died, as Jesus promised he would.

In His sinless and perfect human body, prepared especially as a perfect blood sacrifice for the sins of the world, Jesus suffered terribly in body, soul, and spirit during the long night of His trial - beginning with the agony in the garden of Gethsemane and in all the humiliating events of His trial and cruel torture prior to His morning journey to Golgotha. The worst was yet to come. Death by crucifixion is an especially painful and terrible death.

It was common in Roman times for crucified men in good health to hang dying on a cross sometimes for days, yet Scripture records that Jesus died within six hours' clock time. Even if He only suffered normal human pain in this ordeal it would have been incredibly severe.

All this pain, however, was but the prelude to His real suffering, which involved being cut off from the Father's love and presence and *consigned to carry our sins out of the universe, to hell* as it were, like the scapegoat sacrifice of Israel of which he, Christ, is the antitype.

The Scripture records three statements by Jesus during the first three hours on the cross when He served as the true Great High Priest before the Father and four further statements during the time of darkness from noon to 3 P. M. when the High Priest became the Sin-Offering. It was during the latter three hours, evidently, that the sins of all mankind were laid upon Jesus and the Father turned His face away from His beloved Son. **"For our sake he made him to be sin who knew no sin, so that in him we might become the righteousness of God."** (II Corinthians 5:21)

When contemplating what really took place on the cross in the divine transaction between God the Father and God the Son, we must not think of the sufferings of Christ, terrible as they were (beyond our comprehension), as if they were constrained to a "mere" (endurable) three hours of absolute time. Human beings are basically spirits, and spirits are connected to the eternal dimension. Jesus was not like us in another sense: He had known no sin and suffered the additional revulsion and destruction of being changed from a perfect man into a loathsome, repulsive creature God could not look upon. He *became* sin by absorbing evil into his own person:

**My God, my God, why hast thou forsaken me?**
**Why art thou so far from helping me,**
**from the words of my groaning?**
**O my God, I cry by day, but thou dost not answer;**
**and by night, but find no rest.**
**Yet thou art holy, enthroned on the praises of Israel.**
**In thee our fathers trusted;**
**they trusted, and thou didst deliver them.**
**To thee they cried, and were saved;**
**in thee they trusted, and were not disappointed.**
**But I am a worm, and no man;**

241

scorned by men, and despised by the people.
All who see me mock at me,
they make mouths at me, they wag their heads;
"He committed his cause to the LORD; let him deliver him,
let him rescue him, for he delights in him!"
Yet thou art he who took me from my mother's womb;
thou didst keep me safe upon my mother's breasts.
Upon thee was I cast from my birth,
and since my mother bore me thou hast been my God.
Be not far from me, for trouble is near and there is none to help.
Many bulls encompass me, strong bulls of Bashan surround me;
they open wide their mouths at me,
like a ravening and roaring lion.
I am poured out like water, and all my bones are out of joint;
my heart is like wax, it is melted within my breast;
my strength is dried up like a potsherd,
and my tongue cleaves to my jaws;
thou dost lay me in the dust of the earth.
Yea, dogs are round about me;
a company of evildoers encircle me;
they have pierced my hands and feet -
I can count all my bones--they stare and gloat over me;
they divide my garments among them,
and for my raiment they cast lots.
But thou, O LORD, be not far off! O thou my help, hasten to my aid!
Deliver my soul from the sword, my life from the power of the dog!
Save me from the mouth of the lion,
my afflicted soul from the horns of the wild oxen!
I will tell of thy name to my brethren;
in the midst of the congregation I will praise thee:
You who fear the LORD, praise him!
all you sons of Jacob, glorify him,
and stand in awe of him, all you sons of Israel!
For he has not despised or abhorred the affliction of the afflicted;
and he has not hid his face from him,
but has heard, when he cried to him.
From thee comes my praise in the great congregation;
my vows I will pay before those who fear him.
The afflicted shall eat and be satisfied;
those who seek him shall praise the LORD!
May your hearts live for ever!
All the ends of the earth shall remember and turn to the LORD;

and all the families of the nationsshall worship before him.
For dominion belongs to the LORD, and he rules over the nations.
Yea, to him shall all the proud of the earth bow down;
before him shall bow all who do down to the dust,
and he who cannot keep himself alive.
Posterity shall serve him;
men shall tell of the LORD to the coming generation,
and proclaim his deliverance to a people yet unborn,
that he has wrought it.
(Psalm 22)

Paul wrote many years later of the "fellowship of Christ's sufferings" and of "making up in his own body what is lacking in the sufferings of Christ for the sake of his body, that is the church." He spoke of "always bearing about in the body the dying of Jesus, so that the life of Jesus might be revealed in us...". He said these things long after Jesus had risen from the dead and ascended into heaven where He now rules, His work on the cross having been completed and finished.

Jesus is spoken of in the book of Revelation as the "Lamb slain before the foundation of the world." and Peter writes, **"You know that your were ransomed from the futile ways inherited from your fathers, not with perishable things such as silver or gold, but with the precious blood of Christ, like that of a lamb without blemish or spot. He was destined before the foundation of the world but was made manifest at the end of the times for your sake."** (1 Peter 1:18-20)

Without in any way diminishing the work of Christ on the cross as finished, completed, and accomplished in space-time and in history, it is possible to say that *a part of God suffers eternally* for man's sins. We know from Scripture that God must ultimately remove evil from His presence and bring justice to the world. We also know that those who have permanently rebelled against His gracious mercy do not cease to exist but remain eternally conscious in a place of everlasting, endless punishment. Since God is omnipresent, He, too, is to be found in hell, sustaining its fires and experiencing its pains. In fact, as Bible scholar Ray Stedman once remarked, "Ultimately, God removes evil from the universe by absorbing it into Himself." The so-called "penal view of the atonement" specifies that a Just God must punish sin and that if Christ suffered our punishment as a sustitute He had to bear the full weight of the eternal separation from God that Divine Justice demands and we deserve.

## Yes, There Really is a Hell!

Jesus spoke about hell, giving us this account of an actual happening (not a parable!):

"There was a rich man, who was clothed in purple and fine linen and who feasted sumptuously every day. And at his gate lay a poor man named Lazarus, full of sores, who desired to be fed with what fell from the rich man's table; moreover the dogs came and licked his sores. The poor man died and was carried by the angels to Abraham's bosom. The rich man also died and was buried; and in Hades, being in torment, he lifted up his eyes, and saw Abraham far off and Lazarus in his bosom. And he called out, 'Father Abraham, have mercy upon me, and send Lazarus to dip the end of his finger in water and cool my tongue; for I am in anguish in this flame.' But Abraham said, 'Son, remember that you in your lifetime received your good things, and Lazarus in like manner evil things; but now he is comforted here, and you are in anguish. And besides all this, between us and you a great chasm has been fixed, in order that those who would pass from here to you may not be able, and none may cross from there to us.' And he said, 'Then I beg you, father, to send him to my father's house, for I have five brothers, so that he may warn them, lest they also come into this place of torment.' But Abraham said, 'They have Moses and the prophets; let them hear them.' And he said, 'No, father Abraham; but if some one goes to them from the dead, they will repent.' He said to him, 'If they do not hear Moses and the prophets, neither will they be convinced if some one should rise from the dead'." (Luke 16:19-31)

A similar motif is recorded in the closing verses of Isaiah:
'For as the new heavens and the new earth
which I will make
shall remain before me, says the LORD;
so shall your descendants and your name remain.
From new moon to new moon,
and from sabbath to sabbath,
all flesh shall come to worship me,
says the LORD.

"And they shall go forth and look upon
the dead bodies of the men that have rebelled against me;
for their worm shall not die,
their fire shall not be quenched,
and they shall be an abhorrence to all flesh."
(Isaiah 66:22-24)

## Christ's Return Seen from Eternity

Speaking of the return of Christ as one grand event, Paul writes this to the Thessalonians:

"We are bound to give thanks to God for you, brethren, as is fitting, because your faith is growing abundantly, and the love of every one of you for one another is increasing. Therefore we ourselves boast of you in the churches of God for your steadfastness and faith in all your persecutions and in the afflictions which you are enduring. This is evidence of the righteous judgment of God, that you may be made worthy of the kingdom of God, for which you are suffering - since indeed God deems it just to repay with affliction those who afflict you, and to grant rest with us to you who are afflicted, when the Lord Jesus is revealed (apokalupsis) from heaven with his mighty angels in flaming fire, inflicting vengeance upon those who do not know God and upon those who do not obey the gospel of our Lord Jesus. They shall suffer the punishment of eternal destruction and exclusion from the presence of the Lord and from the glory of his might, when he comes on that day to be glorified in his saints, and to be marvelled at in all who have believed, because our testimony to you was believed. To this end we always pray for you, that our God may make you worthy of his call, and may fulfill every good resolve and work of faith by his power, so that the name of our Lord Jesus may be glorified in you, and you in him, according to the grace of our God and the Lord Jesus Christ." (II Thessalonians 1:3-12)

## What Happens to Those who Reject Jesus?

If death for the follower of Jesus Christ means leaving time and entering eternity, then a similar kind of happening transpires when physical death comes to those who have rejected all of Christ's rights and claims to rule their lives. Since the issue of

245

sin has been dealt with once for all by Jesus, *it is only our proud unwillingness to be forgiven* that ultimately stands between us and our full reconciliation with our Creator! Physical death for those who are not God's children means that spirit, soul and body exit our space-time dimension and "time travel" to the end of the age when the Day of Judgment is held. This event (in eternity) will also intersect human history, like the Second Coming, at some future date on our calendars. But it is no more than a split second away in the consciousness of a person who dies in unbelief! The passage quoted above describes the terrible last glimpse the lost have of Jesus before they are separated from Him forever.

The book of the Revelation records what is known as the judgment of the great white throne, which follows immediatly:

**"Then I (John) saw a great white throne and him who sat upon it; from his presence earth and sky fled away, and no place was found for them. And I saw the dead, great and small, standing before the throne, and books were opened. Also another book was opened, which is the book of life. And the dead were judged by what was written in the books, by what they had done. And the sea gave up the dead in it, Death and Hades gave up the dead in them, and all were judged by what they had done. Then Death and Hades were thrown into the lake of fire. This is the second death, the lake of fire; and if any one's name was not found written in the book of life, he was thrown into the lake of fire."**
(Revelation 20:11-15)

Most Bible commentators believe that all those judged at the great white throne are non-believers, and that there are degrees of punishment in hell, because of the reference to books (angelic records) being opened and the dead being judged according to what they have done. All those present at this judgment will find that their names have *not* been written in the book of life.

## God Suffers Far More than Any Man!

In reading all these passages it is easy to concentrate on what appears to be "unjust" punishment (from our self-righteous human viewpoint), or at least tragic human suffering. We would like to think it could somehow have been avoided! We shrink from thinking too seriously about these passages knowing that but for the grace of God we would have, and should have, been sent to this terrible fate also.

However, a little reflection on the nature of God, for God *is* love, makes it clear that *maintaining hell must be a terrible torment and pain to a God who is self-giving love.* He is the Holy One who "takes no pleasure in the death of the wicked", and who "is not willing that any should perish." God must be willing, therefore, to pay the price of His own eternal pain, suffering, and hell so that the few who are righteous, (by faith), might enjoy eternal bliss. Such is another aspect of the mystery of the suffering of Christ on the cross![5] I think most of us are accustomed to thinking that we suffer more than God, and that He surely cannot fully identify with our minor pains and afflictions.

But a loving God who created us for good things surely *suffers infinitely more* than any human parent when a beloved child refuses the good and chooses the path leading to destruction. Surely it must be grievously painful for a God who is love to be denied the opportunity to give of Himself to the objects of His love. No man can suffer more than Christ has already suffered, nor can mortal man contemplate what is meant by the "longsuffering" of our God (who is outside of time) which will continue, we are assured, at least until the world is changed. **"God's kindness is meant to lead us to repentance,"** Paul writes in Romans, Chapter 2.

## Our God is a Consuming Fire

The letter to the Hebrews says,
**"Therefore let us be grateful for receiving a kingdom that cannot be shaken, and thus let us offer to God acceptable worship with reverence and awe, for *our God is a consuming fire.*"** (Hebrews 12:28,29) The same fires which heal, purify and warm the righteous are the consuming, everlasting burnings of *gehenna*, the garbage dump outside the city, where beings who refused to become the human persons they were designed to be must finally endure the "backside" of love, which is hell. They are discarded because they have not been willing to become (by their own life-long choices) what their Designer intended them to be.

In spite of our outward circumstances and the downward spiral of moral, economic, political, and social conditions over the face of the earth, the Good News of the Bible is that evil has already

247

been dethroned and its power over mankind and nature broken. The new creation is as inevitable as springtime's greening and blossoming.

## Notes to Chapter Eleven

1. Privately published, (Doorway Publications; P.O. Box 291; Brockville, Ontario, Canada; K6V, 5V5, 1980).

2. One of the most wonderful passages on this subject is Isaiah's foreview of the Messiah:

"Behold, my servant shall prosper, he shall be exalted and lifted up, and shall be very high. As many were astonished at him - his appearance was so marred beyond human semblance, and his form beyond that of the sons of men - so shall he startle many nations; kings shall shut their mouths because of him; for that which has not been told them they shall see, and that which they have not heard they shall understand.

"Who has believed what we have heard? And to whom has the arm of the LORD been revealed? For he grew up before him like a young plant, and like a root out of dry ground; he had no form or comeliness that we should look at him, and no beauty that we should desire him. He was despised and rejected by men; a man of sorrows and acquainted with grief; and as one from whom men hide their faces he was despised, and we esteemed him not. Surely he has borne our griefs and carried our sorrows; yet we esteemed him stricken, smitten by God, and afflicted. But he was wounded for our transgressions, he was bruised for our iniquities; upon him was the chastisement that made us whole, and with his stripes we are healed.

"All we like sheep have gone astray; we have turned every one to his own way; and the LORD has laid on him the iniquity of us all. He was oppressed, and he was afflicted, yet he opened not his mouth; like a lamb that is led to the slaughter; and like a sheep that before its shearers is dumb, so he opened not his mouth. By oppression and judgments he was taken away; and as for his generation, who considered that he was cut off out of the land of the living, stricken for the transgression of my people?

"And they made his grave with the wicked and with a rich man in his death, although he had done no violence, and there was no deceit in his mouth. Yet it was the will of the LORD to bruise him; he has put him to grief; when he makes himself an offering for sin, he shall see his offspring, he shall prolong his days; the will of the LORD will prosper in his hand; he shall see the fruit of the travail of his soul and be satisfied; by his knowledge shall the Righteous One, My Servant, make many to be accounted righteous; and he shall bear their iniquities.

"Therefore I will divide him a portion with the great, and he shall divide the spoil with the strong; because he poured out his soul to death, and was numbered with the transgressors; yet he bore the sin of many, and made intercession for the transgressors." (Isaiah 52:13-53:12)

3. R. C. Sproul, *The Holiness of God* (Tyndale House; Wheaton, Illinois, 1985)

4. John R. W. Stott, *The Cross of Christ* (Intervarsity Press; Downers Grove, Illinois, 1986).

5. An excellent discussion about pain and hell is given by C.S. Lewis in his book, *The Problem of Pain* (Macmillan Publishing; New York, 1962). The aspects of time and eternity as they enter into the sufferings of Christ on the cross are also discussed in detail by Arthur Custance in his *Journey out of Time,(op.cit).*

# Chapter Twelve

## Jerusalem Above---the Mother of Us All

Earlier I mentioned participating in a computer conference held in 1984 concerning the planet Mars. During this conference, which lasted approximately one year, the participating scientists discussed, at one point, one member's unusual suggestion. He suggested that perhaps certain pyramidal mountains on Mars were giant housing complexes built by a now-extinct race of giants who lived on the Red Planet. (Since gravity on Mars is much less than on earth we can easily imagine that Martians might by nature be larger than men on earth). Even though our idea about an ancient housing complex on Mars was entirely speculative (the mountains are very likely of natural origin though the mechanism of their formation is unknown), it does no harm to compare the size of one large pyramidal mountain in Cydonia on Mars (near the famous "face on Mars") to the largest pyramid in Egypt. One striking Cydonian Mountain has an estimated volume of about 6000 million cubic meters. If that pyramid were filled with 50 meter cubes (volume of each cube: 125,000 cubic meters), the pyramid would hold 48,000 rooms. A lot of "giant" rooms indeed! Of course, the hypothetical 50-meter cube may only be a structural element. If each cube were further subdivided, we could get many more reasonable sized rooms in a Mars "pyramid". For instance, if an average Martian giant needed a living room 5 meters by 10 by 10 or 500 cubic meters, a Mars pyramidal mountain would hold 10 million rooms! In contrast, the largest of the Egyptian pyramids, that of Cheops at Giza, has a volume of a mere 9 million cubic meters.

After this speculative discussion I called on Bible scholar Dr. Ray Stedman to see whether he had any comments on our fantasy about pyramids on Mars as possible ancient housing complexes. "Did you know", he asked, "that there was one big pyramid mentioned in the Bible?" I knew at once what he was referring to, so I said, "I thought that object was a cube." He replied that, no, he believed it could well be a pyramid. The object in question (a giant, orbiting space station perhaps) is the New Jerusalem, the "city-four-square" described in the next-to-the-last chapter of the Bible. Here is the passage in question:

"Then came one of the seven angels who had the seven bowls full of the seven last plagues, and spoke to me, (John) saying, 'Come, I will show you the Bride, the wife of the Lamb.' And in the Spirit he carried me away to a great, high mountain, and showed me the holy city Jerusalem coming down out of heaven from God, having the glory of God, its radiance like a most rare jewel, like a jasper, clear as crystal. It had a great, high wall, with twelve gates, and at the gates twelve angels, and on the gates the names of the twelve tribes of the sons of Israel were inscribed; on the east three gates, on the north three gates, on the south three gates, and on the west three gates. And the wall of the city had twelve foundations, and on them the twelve names of the twelve apostles of the Lamb. And he who talked to me had a measuring rod of gold to measure the city and its gates and walls. The city lies foursquare, its length the same as its breadth; and he measured the city with his rod, twelve thousand stadia; its length and breadth and height are equal. He also measured its wall, a hundred and forty-four cubits by a man's measure, that is, an angel's. The wall was built of jasper, while the city was pure gold, clear as glass." (Revelation Chapter 21:9-21)

The Book of the Revelation was probably written about 95 A.D. at the latest. Conservative biblical scholarship believes the author to be the same John who wrote the Gospel and three Epistles. We do not have to use much imagination to recognize that John's vision was a great housing complex designed to be inhabited. In the text, 12,000 stadia turns out to be about 1500 statute miles; and if the heavenly city is a pyramid, the pyramid angle is the inverse tangent of the square root of 2, or 54.7 degrees. If every inhabitant were given a cubic mile of real estate in New Jerusalem, the city could accommodate a billion persons comfortably! When the Bible speaks of heaven as the home for the righteous, it is the heavenly city, the New Jerusalem which is meant.

To better understand physical death as a rebirth into a new creation, I have found it helpful to reexamine several very familiar Scriptures in the New Testament that are normally not interpreted in much depth. Before leaving His disciples on the way to the cross Jesus said to them, "Let not your hearts be troubled; believe in God, believe also in me. In my Father's house (the universe) are many rooms; if it were not so, would I have told you

252

that I go to prepare a place for you? And when I go and prepare a place for you, I will come again and will take you to myself, that where I am you may be also." (John 14:1-3)

The place prepared as a future home that Jesus spoke about was evidently the New Jerusalem!

In reply to Rabbi Nicodemus, Jesus spoke of the necessity of being "born from above" in order to enter the kingdom of God (see John, Chapter 3). The experience of spiritual awakening (regeneration), He said, was like physical birth but would be more difficult to understand than physical birth and somewhat mysterious. Christians have always understood spiritual rebirth to be the beginning of a relationship with God which is logically followed by growing from infancy to maturity on the spiritual plane while in this life. Spiritual growth ought to parallel physical growth, we are told. The Bible warns of the dangers of being stuck in spiritual infancy and not growing up spiritually.

Without negating the above interpretation, leaving this present life to enter heaven at the time of physical death can also be thought of, in itself, as a rebirth experience on a grand scale. Indeed, Jesus compared the end of the age we live to childbirth. There would come times of stress at more frequent intervals and each more severe than the last as the age draws to a close. In His farewell meal and discussion with His disciplines Jesus said,

"Truly, truly, I say to you, you will weep and lament, but the world will rejoice; you will be sorrowful, but your sorrow will turn into joy. When a woman is in travail she has sorrow, because her hour has come; but when she is delivered of the child, she no longer remembers the anguish, for joy that a child is born into the world. So you have sorrow now, but I will see you again and your hearts will rejoice, and no one will take your joy from you." (John 16:20-22)

Although it is not biblical to speak of "mother" nature as if she were endued with personality, nevertheless, Paul also uses the figure of birth pangs to describe the close of the age:

"I consider that the sufferings of this present time are not worth comparing with the glory that is to be revealed to us. For the creation waits with eager longing for the revealing of the sons of God; for the creation was subjected to futility, not of its own will but by the will of him who subjected it in hope; because the creation itself will be set free from its bondage to decay and obtain the glorious liberty of the children of God. We know that the whole creation has been groaning in travail together until now; and not only the creation, but we ourselves, who have the first fruits of the Spirit, groan inwardly as we wait for adoption as sons, the redemption of our bodies." (Romans 8:18-23)

A great many Bible commentators have labored over the so-called "intermediate state" of Christians between the time of their physical death and the time of the resurrection of all the Christian dead. Yet the Bible actually teaches us to be absent from the (old) body is to be at home with the Lord. In fact, Paul makes the point that no one wishes to leave his old body behind at death and be found floating in space as a naked spirit while waiting for the raising of the dead. He specifically says, in fact, that our new bodies are already prepared for us and are "put on" at the exact same moment we "put off" our old mortal bodies:

**"For we know that if the earthly tent we live in is destroyed, we have a building of God, a house not made with hands, eternal in the heavens. Here indeed** (in this body) **we groan, and long to put on our heavenly dwelling** (new body), **so that by putting it on we may not be found naked. For while we are still in this tent, we sigh with anxiety; not that we would be unclothed, but that we would be further clothed, so that what is mortal might be swallowed up by life. He who has prepared us for this very thing is God, who has given us the Spirit as a guarantee. So we are always of good courage; we know that while we are at home in the body we are away from the Lord, for we walk by faith, not by sight. We are of good courage, and would rather be away from the body and at home with the Lord. So whether we are at home or away, we make it our aim to please him. For we must all appear before the judgment seat of Christ, so that each one may receive good or evil, according to that he has done in the body."** (II Corinthians 5:1-10)

The problem most of us have in understanding this Scripture is that we carry with us non-biblical notions of time and eternity borrowed from secular science in attempting to understand what happens to a believer at the moment of his departure from this life. The actual situation seems to be that time (as we know it) is measured by the physical bodies we live in, while our spirits are already living in that different time dimension which is characteristic of the heavenly places. Time does not stand still in heaven; heaven is not a condition of timelessness, but a place where time has a different quality and flows at a different rate than time as we experience and measure it on earth in our present fallen condition.

Arthur Custance points out, for example, that throughout the New Testament Jesus is pictured as now being "seated at the right hand of the Majesty on high," whereas at the moment Stephen (the first martyr of the church) was stoned and died, he saw Jesus *standing* at the right hand of God (Acts 7:55). Only when we reach Revelation do we again encounter Jesus "standing" before God in the New Testament record, this time to review the churches, to receive the title deed to the earth, and to begin a new phase of active intervention in human affairs on earth at the close of the age.

While history is flowing on earth according to our familiar calendars and clocks, time and events in the heavenly places are measured by a somewhat different clock and flow of events. And to die and leave this earth is to immediately enter the time-space zone of the heavenlies. I suppose that in "eternity" there is an experience which could be called "timelessness", or the sense that time is frozen or standing still forever. Time sometimes appears to stop for some psychotic persons, and since this is an extremely unpleasant state of mind, I would like to think that having time slow down to an infinitely slow and meaningless pace would be characteristic of hell, not of heaven.

In the experience of the Christian who is dying, the moment of death is accompanied by the subjective experience of leaving time and moving rapidly ("in the twinkling of an eye") ahead to the future to be with Jesus immediately. In experience, he or she perceives the judgment seat of Christ as a rapid succession of experiences and images: the Lord's review and assessment of that person's life and then, a "moment" later, the individual finds himself in the company of all those who are in the process of being raised from the dead and being given their new resurrection bodies. This resurrection is described in I Corinthians 15:35-58:

**"But some one will ask, 'How are the dead raised? With what kind of body do they come?' You foolish man! What you sow does not come to life unless it dies. And what you sow is not the body which is to be, but bare kernel, perhaps of wheat or of some other grain. But God gives it a body as he has chosen, and to each kind of seed its own body. For not all flesh is alike; but there is one kind for men, another for animals, another for birds, and another for fish. There are celestial bodies and there are terrestrial bodies; but the glory of the celestial is one, and the glory of the terrestrial is another. There is one glory of the sun, and another glory of the moon, and another glory of the stars; for star differs from star in glory. So is it with the resurrection of the dead.**

"What is sown is perishable, what is raised is imperishable. It is sown in dishonor, it is raised in glory. It is sown in weakness, it is raised in power. It is sown a physical body, it is raised a spiritual body. If there is a physical body, there is also a spiritual body. Thus it is written, 'The first man Adam became a living being'; the last Adam became a life-giving spirit. But it is not the spiritual which is first but the physical, and then the spiritual. The first man was from the earth, a man of dust, the second man is from heaven. As was the man of dust, so are those who are of the dust; and as is the man of heaven, so are those who are of heaven. Just as we have borne the image of the man of dust, we shall also bear the image of the man of heaven.

"I tell you this, brethren: flesh and blood cannot inherit the kingdom of God, nor does the perishable inherit the imperishable. Lo! I tell you a mystery, We shall not all sleep, but we shall all be changed, in a moment, in the twinkling of an eye, at the last trumpet. For the trumpet will sound, and the dead will be raised imperishable, and we shall be changed.

"For this perishable nature must put on the imperishable, and this mortal nature must put on immortality. When the perishable puts on the imperishable, and this mortal puts on immortality, then shall come to pass the saying that is written: 'Death is swallowed up in victory.' 'O death, where is thy victory? O death, where is thy sting?' The sting of death is sin, and the power of sin is the law. But thanks be to God, who gives us the victory through our Lord Jesus Christ. Therefore, my beloved brethren, be steadfast, immovable, always abounding in the work of the Lord, knowing that in the Lord your labor is not in vain."

No one knows the actual earth-date and earth-time when the resurrection of the dead in Christ will occur. However, we are assured that this event in the future will one day occur as a point of time in earth-history. First Thessalonians, Chapter 4 describes the intersection of the first resurrection with human history. At that "time" there will be Christians (members of the church) alive and on earth. The Lord will return, raising all those in the church who have "died" earlier; that is, those who have previously "fallen asleep." He will then catch up with Him those remaining alive and on earth at the "time" of His return for the church. These living believers will, of course, pass the reviewing process of the judgment seat of Christ and find their old mortal bodies transformed into new resurrection bodies at the same "time."

"But we would not have you ignorant, brethren, concerning those who are asleep, that you may not grieve as others do who have no hope. For since we believe that Jesus died and rose again, even so, through Jesus, God will bring with him those who have fallen asleep. For this we declare to you by the word of the Lord, that we who are alive, who are left until the coming of the Lord shall not precede those who have fallen asleep. For the Lord himself will descend from heaven with a cry of command, with the archangel's call, and with the sound of the trumpet of God. And the dead in Christ will rise first; then we who are alive, who are left shall be caught up together with them in the clouds to meet the Lord in the air; and so shall we always be with the Lord. Therefore comfort one another with these words." (I Thessalonians 4:13-18)

Death is spoken of as "sleep" in several passages of the Bible and so some Christians have suggested that the death of a Christian was followed by a "soul-sleep" - oblivious to the passage of time from the hour of physical death to the resurrection. Other commentators have suggested an intermediate state, as mentioned above, possibly with an intermediate body and a kind of cosmic waiting room where believers stand by for the day the church will be united.

Such ideas can be discarded as unnecessary if we understand earth-time to be relative. The loss felt by those left behind when a loved-one dies is not diminished by the knowledge that they may indeed wait many years to be re-united with the departed. Grief and sorrow and adjustment to great loss are not de-emphasized in Scripture; however, the sting and terror of death have been removed: "Since therefore the children share in flesh and blood, he (Jesus) himself likewise partook of the same nature, that through death he might destroy him who has the power of death, that is, the devil, and deliver all those who through fear of death were subject to lifelong bondage." (Hebrews 2:14,15)

Although Christian writers can and do use allegory as a useful means of helping us to understand spiritual matters, the Bible itself rarely uses the allegorical method of teaching. One exception is found in Paul's letter to the Galatians:

"Tell me, you who desire to be under law (Torah), do you not hear the law? For it written that Abraham had two sons, one by a slave and one by a free woman. But the son of the slave was born according to the flesh, the son of the free woman through promise. Now this is an allegory: these two women are two covenants. One is from Mount Sinai, bearing children for slavery; she is Hagar. Now Hagar corresponds to Mount Sinai in Arabia; she corresponds to the present Jerusalem, for she is in slavery with her children. *But the Jerusalem above is free, and she is our mother.* For it is written, 'Rejoice, O barren one that dost not bear; break forth and shout, you who are not in travail; for the desolate has more children than she who has a husband.' Now we, brothers, like Isaac, are children of promise. But as at that time he who was born according to the flesh persecuted him who was born according to the Spirit, so it is now. But what does the scripture say? 'Cast out the slave and her son; for the son of the slave shall not inherit with the son of the free woman.' So, brothers, we are not children of the slave but of the free woman." (Galatians 4:21-31 My paraphrase)

That New Jerusalem is spoken of as "a bride adorned for her husband" (Revelation 21:2) and in Galatians as "the mother of us all" gives our heavenly home the qualities of redeemed femininity. Just as Eve was called the "mother of all living" and is a type of the church, so the church in her glorified state is described as a woman, as well as a city. In contrast, false religion is also described by the dual figures of a woman and a city (Revelation 17-18). (A warning to avoid this "wrong woman" is given in Proverbs 7:1-27, 8:12-18.)

The creation of man as male and female in the image and likeness of God reveals to us that God is not only a loving Father but also a Mother. A number of writers in our time have noted our difficulty in understanding that God could have feminine, motherly qualities, that He could be intimate and tender and nurturing as well as just and all-powerful and strong and mighty. In fact, in our broken society where so few can identify in a positive way with either parent in many cases, we surely need to discover the new family into which we have been called by the God who created us, redeemed us, and prepared for us a city.

In every sense, the New Jerusalem is our true home, and we arrive there by a process of rebirth into resurrection bodies accompanied by final healing from the devastations of our former sins. We arrive there through the love of the eternal feminine in the Godhead as well as through our heavenly Father's love and the sacrifice of His Son on the cross. We enter our heavenly home at the hour of our death or when Christ returns, but as the writer to the Hebrews says, we are really already there in soul and spirit:

"For you are not come to the mount that might be touched, and that burned with fire, nor to blackness, and darkness, and tempest, And the sound of a trumpet, and a voice whose words made the hearers entreat that no further messages should be spoken to them:

For they could not endure the order that was given, 'If even a beast touches the mountain, it shall be stoned.' Indeed so terrifying was the sight that Moses said, 'I tremble with fear'. But you have come to Mount Zion and to the city of the living God, the heavenly Jerusalem, and to innumerable angels in festal gathering, and to the assembly and of the first born, who are enrolled in heaven, and to a judge who is God of all, and to the spirits of just men made perfect, and to Jesus the mediator of the new covenant, and to the sprinkled blood that speaks more graciously than the blood of Abel. See that you do not refuse him who is speaking. For if they did not escape when they refused him who warned on earth, much less shall we escape, if we turn away from him who warns from heaven. His voice then shook the earth: but now he has promised, saying, 'Yet once more I will shake not only the earth, but also the heaven.' This phrase, 'Yet once more,' indicates the removal of what is shaken, as of things that are made, in order that what cannot be shaken may remain. Therefore let us be grateful for receiving a kingdom which cannot be shaken, and let us offer to God acceptable worship, with reverence and awe; for our God is a consuming fire." (Hebrews 13:18-29) (Paraphrased KJV)

The subject of the last seven years of the age in which we live has been dealt with so frequently by popular Christian writers of our day that even many non–Christians are acquainted with the scenario of the Last World War and Second Coming of Christ in Israel at the Battle of Armageddon. This open and visible return of Christ *with* His saints (as opposed to His earlier silent return *for* His saints described above) will usher in a time of healing and restoration for a world ruined by famine, war, plague, nuclear war and nuclear winter. Then ancient promises to the nation of Israel will be fulfilled and nations will be healed. Whether or not the laws of physics as discussed will be changed at this time, or later, at the end of the thousand-year reign of Christ on earth, I do not know. The coming again of Jesus in history is, however, as certain as the fact that we all must face Him when we die whether we know Him or not. The eventual transformation of the entire universe is already guaranteed:

"For behold, I create new heavens and a new earth;
and the former things shall not be remembered
or come into mind.
But be glad and rejoice for ever
in that which I create;

for behold, I create Jerusalem a rejoicing,
and her people a joy.
I will rejoice in Jerusalem,
and be glad in my people;
no more shall be heard in it the sound of weeping
and the cry of distress.
No more shall there be in it
an infant that lives but a few days,
or an old man who does not fill out his days,
for the child shall die a hundred years old,
and the sinner a hundred years old
shall be accursed.
They shall build houses and inhabit them;
they shall plant vineyards and eat their fruit.
They shall not build and another inhabit;
They shall not plant and another eat;
for like the days of a tree shall the
days of my people be,
and my chosen shall long enjoy the
work of their hands.
They shall not labor in vain,
or bear children for calamity;
for they shall be the offspring of the
blessed of the LORD,
and their children with them.
Before they call I will answer,
while they are yet speaking I will hear.
The wolf and the lamb shall feed together,
the lion shall eat straw like the ox;
and dust shall be the serpent's food.
They shall not hurt or destroy
in all my holy mountain, says the Lord."
(Isaiah 65:17-25)

# Appendix

## What the Bible Says About Itself

"All scripture is inspired by God and profitable for teaching, for reproof, for correction, and for training in righteousness, that the man of God may be complete, equipped for every good work." (II Timothy 3:16)

"Is not my word like fire, says the LORD, and like a hammer which breaks the rock in pieces?" (Jeremiah 23:29)

"Then they cried to the LORD in their trouble, and he delivered them from their distress;  he sent forth his word and healed them, and delivered them from destruction." (Psalm 107:19-20)

"How can a young man keep his way pure?  By guarding it according to thy word.  With my whole heart I seek thee; let me not wander from your commandments!  I have laid up thy word in my heart, that I might not sin against thee." (Psalm 119:9-11)

"You have been born anew, not of perishable seed but of imperishable, through the living and abiding word of God; for, 'All flesh is like grass and all its glory like the flower of the grass. The grass withers, and the flower falls, but the word of the LORD abides forever.' That word is the good news which was preached to you." (I Peter 1:23-25)

"Heaven and earth will pass away, but my words will not pass away." (Matthew 24:35)

"For the word of God is living and active, sharper than any two-edged sword, piercing to the division of soul and spirit, of joints and marrow, and discerning the thoughts and intentions of the heart." (Hebrews 4:12)

"And take...the sword of the Spirit which is the word of God." (Ephesians 6:17)

"Every word of God proves true.  He is a shield to all those who take refuge in him." (Proverbs 30:5)

"For my thoughts are not your thoughts, neither are your ways my ways, says the LORD. For as the heavens are higher than the earth, so are my ways higher than your ways and my thoughts than your thoughts. For as the rain and snow come down from heaven and return not thither but water the earth, making it bring forth and sprout, giving seed to the sower and bread to the eater, so shall my word be that goes forth from my mouth; it shall not return to me empty, but it shall accomplish that which I purpose, and prosper in the thing for which I sent it." (Isaiah 55:8-11)

"Study to show thyself approved unto God, a workman who needeth not to be ashamed, rightly dividing the word of truth."
( 2 Timothy 2:15)

"And we also thank God constantly for this, that when you received the word of God which you heard from us, you accepted it not as the word of men but as what it really is, the word of God, which is at work in you believers." (I Thessalonians 2:13)

"For the word of the LORD is upright, and all his work is done in faithfulness. He loves righteousness and justice; the earth is full of the steadfast love of the LORD. By the word of the LORD the heavens were made, and all their host by the breath of his mouth. He gathered the waters of the sea as in a bottle; he put the deeps in storehouses. Let all the earth fear the LORD, let all the inhabitants of the world stand in awe of him! For he spoke, and it came to be; he commanded, and it stood forth." (Psalm 33:4-9)

## What the Old Testament Says About the Messiah

The First promise of a Redeemer: God speaking to Eve in the Garden, "I will put enmity between you and the woman, and between your seed and her seed; he shall bruise your head, and you shall bruise his heel." (Genesis 3:15)

Messiah to be in the line of Judah. Jacob's prediction: "You are a lion's cub, O Judah: you shall return from the prey my son. Like a lion he crouches and lies down, like a lioness - who dares to rouse him? The scepter will not depart from Judah, nor the ruler's staff from between his feet, until he (Messiah) comes to whom it belongs, and the obedience of the nations is his. He will tether his donkey to a vine, his colt to the choicest branch; he will wash his garments in wine, his robes in the blood of

grapes. His eyes will be darker than wine, his teeth whiter than milk."
(Genesis 49:9-11) (Paraphrased KJV)

The line of Seth, Genesis 4-5; The line of Shem, Genesis 11:10-26.
Messiah in the line of Abraham, Isaac, and Jacob, Genesis 12ff.

A prophet to be raised up like Moses: "The LORD your God will raise up for
you a prophet like me (Moses) from among you, from your brethren -
him you shall heed - just as you desired of the LORD your God at
Horeb on the day of the assembly, when you said, 'Let me not hear
again the voice of the Lord my God, or see this great fire any more, lest
I die.' And the LORD said to me, 'They have rightly said all that they
have spoken. I will raise up for them a prophet like you from among
their brethren; and I will put my words in his mouth, and he shall
speak to them all that I command him. And whoever will not give heed
to my words which he shall speak in my name, I myself will require it
of him.'" (Deuteronomy 18:15-19)

Job's Testimony: "For I know that my Redeemer lives, and at last he will
stand upon the earth; and after my skin has been thus destroyed, then
from (without) my flesh I shall see God, whom I shall see on my side,
and my eyes shall behold, and not another..." (Job 19:25-27)

Balaam's Fourth Blessing and Prophecy: "...The oracle of Balaam the son of
Beor, the oracle of the man whose eye is opened, the oracle of him who
hears the words of God, and knows the knowledge of the Most High,
who sees the vision of the Almighty, falling down, but having his eyes
uncovered: I see him, but not now; I behold him, but not nigh: a star
shall come forth out of Jacob, and a scepter shall rise out of Israel; it
shall crush the forehead of Moab, and break down all the sons of
Sheth." (Numbers 24:15-17)

Messiah as the Servant of Yahweh: "Behold, my servant, whom I uphold,
my chosen, in whom my soul delights. I have put my Spirit upon him,
he will bring forth justice to the nations. He will not cry or lift up his
voice, or make it heard in the street; a bruised reed he will not break,
and a dimly burning wick he will not quench; he will faithfully bring
forth justice. He will not fail or be discouraged till he has established
justice in the earth; and the coastlands wait for his law." (Isaiah 42:1-4)

The Branch out of Jesse: "There shall come forth a shoot from the stump
of Jesse, and a branch shall grow out of his roots. And the Spirit of the
Lord shall rest upon him...And his delight shall be in the fear of the
LORD. He shall not judge by what his eyes see, or decide by what his
ears hear; but with righteousness he shall judge the poor, and decide

with equity for the meek of the earth; and he shall smite the earth with the rod of his mouth, and with the breath of his lips he shall slay the wicked. Righteousness shall be the girdle of his waist, and faithfulness the girdle of his loins...In that day the root of Jesse shall stand as an ensign to the peoples; him shall the nations seek, and his dwellings shall be glorious." (Isaiah 11:1-5,10)

Permanency of the line of King David, and of the Land of Israel:
2 Samuel 7:11ff; 2 Samuel 23; Ezekiel 37:24-28; Jeremiah 30-31.

The work of Messiah: "The Spirit of the LORD God is upon me (Messiah), because the LORD has anointed me to bring good tidings to the afflicted; he has sent me to bind up the brokenhearted, to proclaim liberty to the captives, and the opening of the prison to those who are bound; to proclaim the year of the LORD's favor, and the day of vengeance of our God; to comfort all who mourn; to grant to those who mourn in Zion - to give them a garland instead of ashes, the oil of gladness instead of mourning, the mantle of praise instead of a faint spirit; that they might be called oaks of righteousness, the planting of the LORD, that he may be glorified." (Isaiah 61:1-3)

Messiah's Fiery Coming: "Who is this that comes from Edom, in crimsoned garments from Bozrah, he that is glorious in his apparel, marching in the greatness of his strength? 'It is I, announcing vindication, mighty to save.' Why is thy apparel red, and thy garments like his that treads in the wine press? 'I have trodden the wine press alone, and from the peoples no one was with me; I trod them in my anger, and trampled them in my wrath, their lifeblood is sprinkled upon my garments, and I have stained all my raiment. For the day of vengeance was in my heart, and my year of redemption has come. I looked, but there was no one to help; I was appalled, but there was no one to uphold; so my own arm brought me victory, and my wrath upheld me. I trod down the peoples in my anger, I made them drunk in my wrath, and I poured out their lifeblood upon the earth."
(Isaiah 63:1-6)

Messiah's Constant Availability: " I was ready to be sought by those who did not ask for me; I was ready to be found by those who did not seek me. I said, 'Here am I, here am I,' to a nation that did not call on my name. I spread out my hands all the day to a rebellious people, who walk in a way that is not good, following their own devices."
(Isaiah 65:1,2)

Messiah, the Branch and His future reign over Israel: "In that day the branch of the LORD shall be beautiful and glorious, and the fruit of the land shall be the pride and glory of the survivors of Israel. And he who is left in Zion and remains in Jerusalem will be called holy, every one who has been recorded for life in Jerusalem, when the LORD shall have washed away the filth of the daughters of Zion and cleansed the bloodstains of Jerusalem from its midst by a spirit of judgment and by a spirit of burning. Then the LORD will create over the whole site of Mount Zion and over her assemblies a cloud by day, and smoke and the shining of a flaming fire by night; for over all the glory there will be a canopy and a pavilion. It will be for a shade by day from the heat, and for a refuge and a shelter from the storm and rain." (Isaiah 4:2-6)

Messiah to be born of a virgin: "Therefore the LORD himself will give you a sign. Behold a young woman shall conceive and bear a son and you shall call his name Immanuel." (Isaiah 7:14)

Birth of the Messiah as King of Israel: "For to us a child is born, to us a son is given; and the government will be upon his shoulder, and his name will be called 'Wonderful Counselor, Mighty God, Everlasting Father, Prince of Peace.' Of the increase of his government and of peace there will be no end, upon the throne of David, and over his kingdom, to establish it, and to uphold it with justice and with righteousness from this time forth and for evermore. The zeal of the LORD of hosts will do this." (Isaiah 9:6-7)

Safety and Fulfillment for Israel under Messiah: "Your eyes will see the king in his beauty; they will behold a land that stretches afar...Look upon Zion, the city of our appointed feasts! Your eyes will see Jerusalem, a quiet habitation, an immovable tent, whose stakes will never be plucked up, nor will any of its cords be broken. But there the LORD in majesty will be for us a place of broad rivers and streams, where no galley with oars can go, nor stately ship can pass. For the LORD is our judge, the LORD is our ruler, the LORD is our king; he will save us." (Isaiah 33:17,20-22)

Future Restoration of Israel under Messiah: "For the children of Israel shall dwell many days without king or prince, without sacrifice or pillar, without ephod or teraphim. Afterward the children of Israel shall return and seek the LORD their God, and David their king; and they shall come in fear to the LORD and to his goodness in the latter days." Hosea 3:4,5)

Messiah from Bethlehem: "But you, O Bethlehem Ephrathah, who are little to be among the clans of Judah, from you shall come forth for me one who is to be the ruler in Israel, whose origin is from of old, from ancient days...And he shall stand and feed his flock in the strength of the LORD, in the majesty of the name of the LORD his God. And they shall dwell secure, for now he shall be great to the ends of the earth." (Micah 5:2,5)

Messiah to Ride into Jerusalem on a donkey: "Rejoice greatly, O daughter of Zion! Shout aloud, O daughter of Jerusalem! Lo, your king comes to you; triumphant and victorious is he, humble and riding on an ass, on a colt the foal of an ass...and he shall command peace to the nations; his dominion shall be from sea to sea, and from the River (Euphrates) to the ends of the earth." (Zechariah 9:9-10)

Coming of Messiah to the Mount of Olives: "On that day his feet shall stand on the Mount of Olives which lies before Jerusalem on the east; and the Mount of Olives shall be split in two from east to west by a very wide valley; so that one half of the Mount shall withdraw northward, and the other half southward...Then the LORD your God will come, and all the holy ones with him...And the LORD will become king over all the earth; on that day the LORD will be one and his name one." (Zechariah 14:4-9)

Israel to look upon Messiah who was pierced: "And I will pour out on the house of David and the inhabitants of Jerusalem a spirit of compassion and supplication, so that, when they look on him whom they have pierced, they shall mourn for him, as one mourns for an only child, and weep bitterly over him, as one weeps over a first-born." (Zechariah 12:10,11)

Rejection of Messiah as Good Shepherd and the false shepherd, Zechariah 11:4ff. Messiah betrayed at the house of his friends, Zechariah 13:6.

Messiah the Righteous Branch of David: "Behold, the days are coming says the LORD, when I will raise up for David a righteous branch, and he shall reign as king and deal wisely, and shall execute justice and righteousness in the land. In his days Judah will be saved, and Israel will dwell securely. And this is the name by which he will be called: 'The LORD is our Righteousness.'" (Jeremiah 23:5,6)

The Branch, Both King and High Priest of Israel: "...behold, I will bring my servant the Branch...and I will remove the guilt of this land in a single day. In that day, says the LORD of hosts, every one of you will invite his neighbor under the vine and under his fig tree." (Zechariah 3:8-10) "Behold, the man whose name is the Branch: for he shall grow up in his place, and he shall build the temple of the LORD. It is he who shall bear royal honor, and shall sit and rule upon his throne. And there shall be a priest by his throne, and peaceful understanding shall be between them both." (Zechariah 6:12,13)

Messiah, the son of man: "As I looked, thrones were placed and one that was ancient of days took his seat; his raiment was white as snow, and the hair of his head like pure wool; his throne was fiery flames, its wheels were burning fire. A stream of fire issued and came forth from before him; a thousand thousands served him, and ten thousand times ten thousand stood before him; the court sat in judgment, and the books were opened...I saw in the night visions and behold, with the clouds of heaven there came one like a son of man, and he came to the Ancient of Days and was presented before him. And to him was given dominion and glory and kingdom, that all peoples, nations, and languages should serve him; his dominion is an everlasting dominion, which shall not pass away, and his kingdom one that shall not be destroyed." (Daniel 7:9,10; 13,14)

# The Psalms and the Messiah

The Messianic Psalms: 2, 8, 16, 22, 23, 24, 40, 41, 45, 68, 69, 72, 89, 102, 110 , 118.
Messiah the Coming King,  Messiah to be raised from the dead, Psalm 16.
Sufferings of the Messiah, His death, and His triumph which follows, Psalm 22.
Messiah as the Good Shepherd, Psalm 23.
The Ascension of the King of Glory, Psalm 24 .
Messiah's Dependence upon the Father and His servant character, Psalm 40.
Slanders against God's Messiah, His betrayal, Psalm 41.
Messiah and His Bride, Psalm 45.
Israel under Messiah's Protection and guidance, Her future glory, Psalm 68.
Prayer for deliverance, imprecation for enemies of God, and praise,  Psalm 69.
Messiah as ideal King, and His glorious kingdom, Psalm 72.
The Covenant with David, fulfilled  through Messiah, Psalm 89.
Messiah's Cry in Affliction, Psalm 102.
Messiah's kingly rule over His enemies; high priesthood, Psalm 110.
The loving-kindness of God shown forth in Messiah, Psalm 118.

# The Use of Numbers and Symbols In the Bible

In reading the Bible, it soon becomes apparent that certain numbers (such as 3, 7, 12, 40, etc) occur more frequently than we would expect. From the context it is soon apparent that some of these numbers have symbolic as well as literal meaning. This is not the same thing as assigning numerical values to the Greek and Hebrew letters that comprise the text. The latter approach does allow useful computer analysis of the structure of the Bible from which authorship of various books can be confirmed and certain obscure passages (manuscript errors) clarified.

The number **ONE** stands for unity. "There is ONE God, and ONE mediator between God and men, the man Christ Jesus, who gave himself a ransom for many." A second example is the S'hma, "Hear O Israel the LORD your God is ONE God..."

**TWO** stands for union. "For this cause (marriage) a man shall leave his father and his mother and the TWO shall become one flesh." Man, created as Adam/Eve, was separated into Adam and Eve so the two could complement one another in a new kind of unity. Basis of the BINARY system of counting.

**THREE** is the number of God: Father, Son, and Holy Spirit. The THIRD Day is the day of resurrection in scripture.

**FOUR** is the number of the world system (cosmos). Thus, we have the FOUR winds, the FOUR seasons, the FOUR corners of the earth, and the FOUR living creatures (angels) around the throne of God.

**FIVE** is the number of division. Thus, there were FIVE wise and FIVE foolish virgins invited to the marriage feast.

**SIX** is the number of man because man was created on the SIXTH day. The number 666 is the number of the antichrist, the man who proclaims himself to be God. SIX days are appointed for man to work. The seventh day (the Sabbath) is to be devoted to rest.

**SEVEN** is the number of completeness in the old created order, (4+3=7). There are seven days of creation and in Revelation this number occurs 54 times.

**EIGHT** is the number of new beginnings. Jesus rose from the dead on the Eighth Day of the week, a Sunday. Basis of the OCTAL system of counting.

**TEN** is the number of worldly completion (4+6=10). Ten virgins total. Ten toes on the image in Nebuchadnezzar's dream, ten horns on the beast of Revelation 17:2. Basis of the DECIMAL System.

**TWELVE** is the number of eternal perfection as in the TWELVE Tribes of Israel and the TWELVE apostles of the church (12+12=24 elders). There are TWELVE gates to New Jerusalem, its walls measure 12 X 12 cubits high and the sides are 12,000 furlongs in length.

**FORTY** is the number of testing. Jesus was tempted in the wilderness FORTY days and nights. Israel wandered in the wilderness FORTY years. During the flood of Noah, the rain fell FORTY days and nights.

**FIFTY** is the number of the Jubilee Year in Israel. The land was to rest every seventh year and after 7X7+1=50 years to revert to original owners with debts being forgiven. Pentecost occurred seven sabbaths plus one day after the death of Jesus, on a Sunday morning. **SEVENTY** is the number of years the Jews spent out of the land during the Babylonian captivity. SEVENTY "weeks of years" were given to Israel (See Daniel) for the completion of their national destiny.

## Some Common Symbols in the Bible

**METALS:** Various metals are mentioned in the Bible - including gold, silver, brass, iron, lead, and tin. **GOLD** is a symbol of deity prominent in the Tabernacle of Moses and the First and Second Temples. **SILVER** stands for redemption and was used in quantity in the tabernacle. **BRASS** stands for judgment, as in the altar of sacrifice, or Moses' Serpent of Brass. **LEAD** and **TIN** are baser metals removed by refining as dross to yield silver and gold. **IRON** stands for military (or industrial strength), and IRON mingled with **CLAY** (Daniel) represents the weakness of modern Western democracies which combine rule by the common people with military and industrial strength. **WOOD** represents humanity.

**COLORS: SCARLET** represents natural life in man since "the life is in the blood." This is the color of "life poured out" to ransom us. **WHITE** stands for purity; **BLUE,** for the heavens; **BLACK,** for death; **GREEN,** for new life in nature; **PURPLE,** for royalty and **PALE GREEN,** for famine. **GARMENTS** stand for righteousness. Imputed faith is represented by white garments and filthy rags represent self-righteousness, or defilement by the world.

**HONEY** represents natural sweetness (a characteristic of the flesh; hence, not a positive symbol). **LEAVEN** (yeast) is a universal symbol of sin in scripture.

**ANIMALS: HORSES** are always "war horses" (imported from Egypt) and represent reliance on the world's resources rather than on God in spiritual warfare. **DOGS** are unclean animals and refer to the lowest forms of behavior by men. **WOLVES** are false teachers who prey upon the **SHEEP.** The latter represent believers who are helpless, dumb, prone to wander, and always in need of a shepherd. **BULLS** and **RAMS** represent masculine strength. **LIONS** and **LEOPARDS** can represent spiritual enemies in the heavenly places ("...your adversary the devil goes around like a prowling lion..."). On the other hand, Jesus is the "Lion of the Tribe of Judah." **GAZELLES, HINDS** and **DEER** represent youthful energy and vitality as seen in the Song of Solomon. The **BEASTS** in Revelation (and Daniel) are men (world leaders) as seen by God, except that the "four living creatures" in Revelation are mighty angels known as cherubim.

269

BIRDS: The **DOVE** is a symbol of the Holy Spirit. The **EAGLE** pictures the serene sovereignty of God. (God took Israel out of Egypt "on eagles' wings". **BIRDS** of **PREY** represent forces of destructiveness. Animal **HORNS** picture power in a ruler.

The **SEA** stands for the masses of mankind. The "great sea" is the Mediterranean and the "eastern sea" or "salt sea" is the Dead Sea. **MOUNTAINS** are symbols of human government. The **STARS** frequently represent the angels. The **SUN** is a symbol for Christ who rules the earth by day; and the **MOON**, the church who rules the earth by night", and has no light of its own.

The **NECK** symbolizes the will (as in "stubborn and stiff-necked"). The **LOINS** represent masculine virility and strength. The **FEET** indicate activity: "How beautiful on the mountains are the feet of those who preach the gospel..." or, "Their feet are swift to shed blood." The **EYES** represent spiritual insight and inner beauty. A **KISS** is a symbol for intimacy.

**INCENSE** symbolizes prayer; and **MYRHH**, suffering. **FAT** stands for the natural richness of life; hence, the fat of the sacrifices belongs to God alone.

**GRASS** represents the ephemeral life of man. Good **FRUIT** is the result of living in dependence on Christ, and bad fruit, the natural result of self-life. **THORNS** and **THISTLES** stand for the works of the flesh.

**MILK** is spiritual food for new-born babes in Christ; **BREAD** for growing young men, and the **MEAT** of the Word, deeper truth for the maturing believer.

A **WOMAN** typifies every believer in such passages as Romans 7:1-4. A **HARLOT** symbolizes fallen mankind (men as well as women), and the **GREAT HARLOT,** the apostate church and the final state of the world system under the dominion of evil. The true church and the false church are both symbolized by the figures of a woman and by a city.

The **VINE** represents Israel's national influence among the nations. The **FIG TREE** stands for Israel's religious history; and the **OLIVE TREE**, Israel's true spiritual history apart from outward religious appearances.

The HOLY SPIRIT is symbolized in Scripture by (a) a dove, (b) water, (c) oil, (d) fire, (e) wind or breath, (f) light.

Without denying the historical validity of the Old Testament in any way, CROSSING THE RED SEA is a picture of leaving the world (Egypt) under the domain of the god of the age (Pharaoh) to be baptised into Christ at the Red Sea crossing. The WILDERNESS stands for Christian living under the law in the power of the flesh. CROSSING THE JORDAN stands for the truth of Galatians 2:20, entering the land, (the Spirit-filled life); by renouncing self and living thereafter in the power of the Spirit. The ENEMIES in the LAND: CanaanITES, HivITES, JebusITES, EdomITES, MoabITES are pictures of the flesh. JACOB represents the average believer, deceiving and deceived. ISRAEL is the man who wrestled with God (*the* Angel of the LORD), and becomes changed as a result, a picture for us of sanctifcation. ESAU, the brother of JACOB is the father of the EDOMITES, hence a picture of the flesh. Agag, Haman, and Herod are notable among the descendants of Esau. ASSYRIA represents lawlessness, BABYLON, religious confusion and error. The Old Testament pictures every believer as a king over the kingdom of his (or her) life. Power to rule in righteousness is given only as we subject ourselves to the King of kings, Jesus.

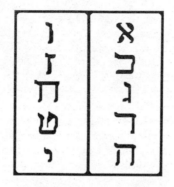